Underwater World

TIME-LIFE
ALEXANDRIA, VIRGINIA

CONTENTS

4 Desperately Seeking Sustenance

5 The Art of Self-Defense

1

All Creatures Wet and Wild

From clear mountain streams to muddy tropical swamps, from small, shallow ponds to vast, deep seas, the waters of the world teem with a wondrous variety of creatures. The oceans alone contain more than 13,000 kinds of fish. Another 8,400 fish species live in fresh water, and a select few—salmon and eels, for example—are able to commute between the two environments. The realm of water inhabitants also includes mollusks, such as clams and snails, crustaceans, such as crabs and lobsters, and a rich array of mammals, reptiles, and birds.

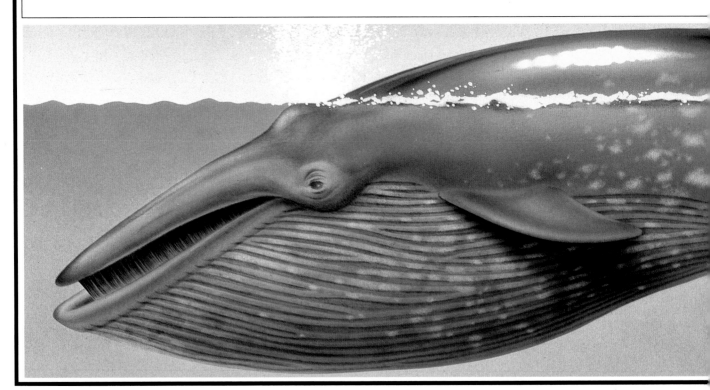

Not only do aquatic creatures come in a remarkable range of shapes and sizes, but each animal is uniquely suited to its habitat. Some of them swim freely, while others float or drift with the currents. Some crawl or rest on the seafloor or riverbeds; some cling to rocks; and still others bury themselves in the sand.

Although the tropical and temperate zones of the oceans provide a relatively stable environment, other watery worlds are far less hospitable. Some rivers, for example, dry up and disappear during periods of drought. Harsh conditions also prevail in the polar seas, where the water temperature rarely exceeds 30° F. Even in these icy waters, however, some creatures manage to survive—and thrive. How fish and other water-dwelling animals have adapted to their surroundings is the subject of this chapter.

Symbolizing the wide range of creatures that inhabit the oceans, the ubiquitous jellyfish *(above)* and the rare blue whale *(below)* both move in a mode tailored to their shape: The jellyfish expands and contracts its body, while the whale swings its tail fin up and down.

What Is a Fish?

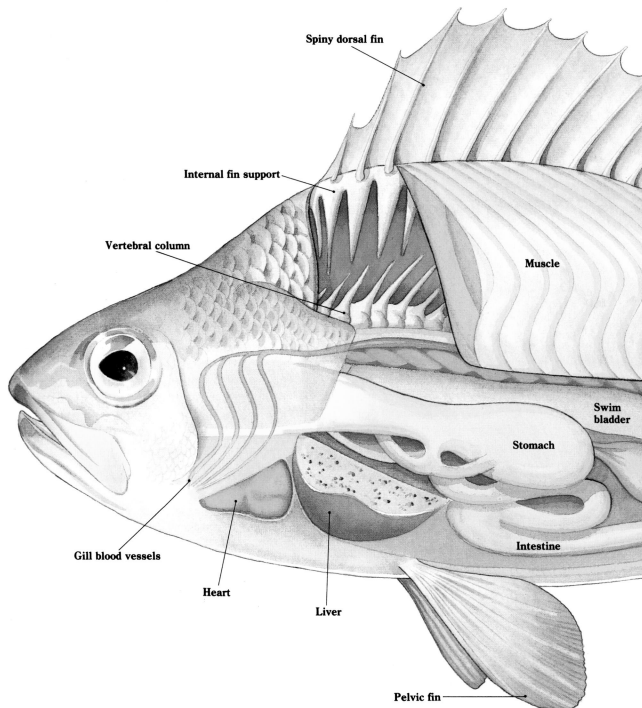

Spiny dorsal fin

Internal fin support

Vertebral column

Muscle

Swim bladder

Stomach

Intestine

Gill blood vessels

Heart

Liver

Pelvic fin

The three main types of fishes

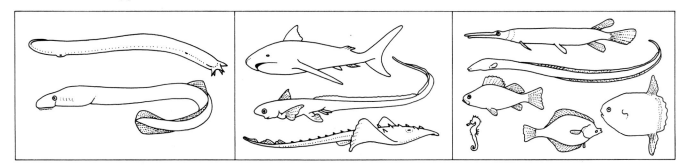

The jawless fishes, which include lampreys and hagfishes, are the most primitive living vertebrates.

The cartilaginous fishes— including sharks, rays, and skates—have skeletons made of cartilage, not bone.

Bony fishes, the largest class of vertebrates, have skeletons made of bone. Most of these fishes are covered with scales.

Scientists define a fish as a cold-blooded aquatic vertebrate. In plainer terms, this means its blood temperature matches that of the surroundings, it lives in water, and it has a backbone. Most fish breathe with gills, reproduce by laying eggs, and are covered with protective scales.

A number of specialized physical features make fish well adapted to aquatic life. Because their natural buoyancy makes them nearly weightless in water, fish have developed light, flexible skeletons. Attached to this framework are strong muscles, such as those of the perch shown below, that power the fish through the water. Fins—winglike structures on the back, belly, sides, or tail—act as rudders, stabilizers, or paddles. The internal organs, including the all-important swim bladder *(pages 12-13),* are in the body cavity beneath the vertebral column.

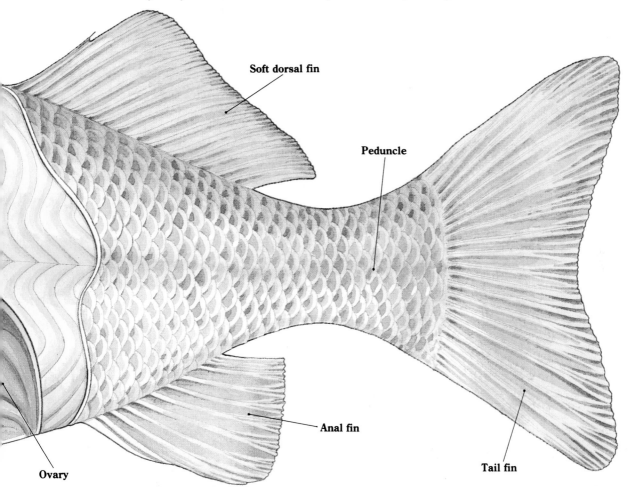

Soft dorsal fin

Peduncle

Anal fin

Tail fin

Ovary

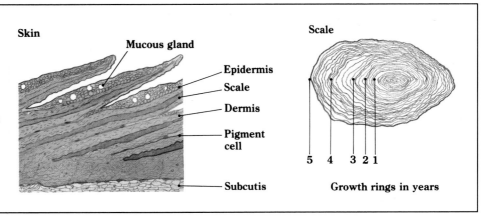

The skin of a fish has two layers: the outer epidermis and the dermis below it. Mucus, secreted by glands in the epidermis, protects the skin against fungi and bacteria. Scales, hardier protection, are made of transparent bone embedded in the dermis. Like tree rings, they provide a record of age and growth.

Skin

Mucous gland

Epidermis

Scale

Dermis

Pigment cell

Subcutis

Scale

5 4 3 2 1

Growth rings in years

How Do Fish Breathe Underwater?

Like almost all living creatures, fish need oxygen in order to exist. Most of them obtain this vital element through special sievelike organs that are called gills.

Located just behind the mouth cavity on both sides of the head, the gills are usually covered by a flap known as the operculum. Underneath the operculum are four overlapping rows of blood-red gills. Each gill consists of a bony arch that supports numerous gill filaments—pairs of slender, fleshy appendages that resemble closely spaced teeth on a comb. Protruding from each filament are tiny membranes, or lamellae, that contain myriad blood capillaries. The walls of the lamellae are so thin that blood flowing through them extracts oxygen from water passing across the gills. As part of the same process, the lamellae remove carbon dioxide from the blood and expel it into the water. Water contains 1/30 as much oxygen as air does, and this gas exchange—oxygen in, carbon dioxide out—is a key component of aquatic survival.

Getting water to the gills

To get enough oxygen to live, most fish must receive a constant supply of aerated water to the gills. In many bony fish, the mouth and gills work together as a suction pump: First, the gill covers close while the mouth opens and its walls expand, drawing water inward. Second, the mouth cavity contracts and the mouth shuts while the gill covers open, forcing the water out of the mouth and across the gills. This method of respiration, which enables water to reach the gills even when the fish is not swimming about, is found in sedentary fish such as carp, flounder, and sole.

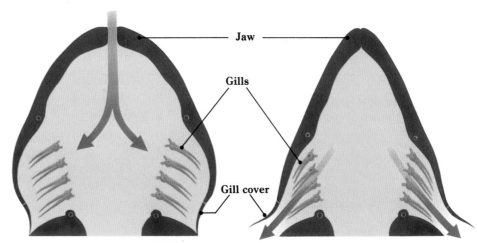

Breathing begins when the fish's mouth opens and its oral cavity expands, sucking in water *(arrows)*.

Next, the fish's mouth closes and the gill covers open, pumping water over the gills and out of the gill cavity.

Gill
raker

Gill filament

Gill arch
(branchial
cartilage)

Blood vessel

Blood vessel

Stiff gill rakers on the gill arch filter incoming water. Blood vessels in the gill filaments supply and drain capillaries in the lamellae.

Blood
vessel

Water flow

Blood flow

Lamella

Blood vessel

As water flows over a gill filament, arterial vessels draw oxygen from the water. Venous vessels carry blood loaded with carbon dioxide to the lamellae, where it is discharged.

True mouth breathers

Very active fish—mackerel, tuna, and some sharks—need more oxygen than sluggish species such as catfish, flounder, stingrays, eels, and sea horses. This is why active fish often swim with their mouths open; with a greater volume of water passing across the gills, oxygen intake rises. In addition, the gills of these species tend to have larger surface areas and thinner, more closely spaced lamellae; this too makes for maximum oxygen absorption. Such fish must keep swimming, however, even when they are asleep; if not, they will die by suffocation.

Active fish like the bluefin tuna *(above)* swim with mouths agape.

Do Fish Drink Water?

The amount of water a marine fish imbibes depends on the salinity of its environment. The saltier the water, the more the fish drinks.

Saltwater, or marine, fish are constantly losing water and absorbing salt. To prevent dehydration, they drink from 7 to 35 percent of their body weight in seawater every day.

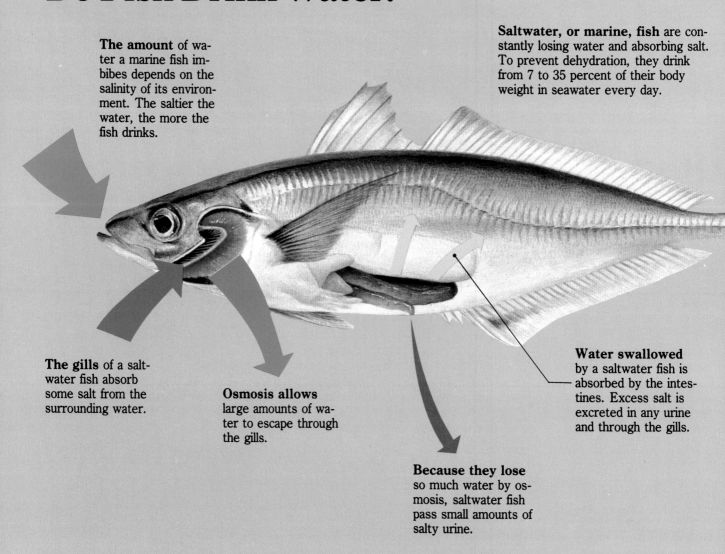

The gills of a saltwater fish absorb some salt from the surrounding water.

Osmosis allows large amounts of water to escape through the gills.

Because they lose so much water by osmosis, saltwater fish pass small amounts of salty urine.

Water swallowed by a saltwater fish is absorbed by the intestines. Excess salt is excreted in any urine and through the gills.

Trading places

Most marine fish would perish if transferred to fresh water, just as most freshwater fish would die if moved to salt water. Out of their normal environments, both types of fish lack the ability to adjust the salt density in their bodies.

When a saltwater fish is placed in fresh water, the water that usually flows outward through its body reverses direction. The creature swells up and soon dies. And when a freshwater fish is put in seawater, the water in its tissues flows out toward the stronger saline solution of the ocean. As a result, the concentration of salt in its body fluids rises, killing the fish.

In its usual home, a saltwater fish maintains the proper saline balance of its body fluids by drinking seawater and excreting excess salt.

In fresh water, the marine fish absorbs water, diluting its body fluids. Unable to retain salt in its body or get rid of the excess water, the fish dies.

Normally, a freshwater fish regulates the salinity of its tissues by absorbing salt and eliminating water.

In salt water, the fish loses water it cannot replace; its salt content rises to a lethal level.

Although saltwater fish drink great quantities of water, freshwater fish drink almost none at all. The difference stems from the creatures' need to maintain the proper balance of salt and water in their bodies.

The oceans are three times saltier than the body fluids of the fish who live there. Because of the natural process known as osmosis, the water inside a saltwater fish tends to flow through its skin and gills into the seawater outside. To replace this lost liquid, the saltwater fish must swallow huge amounts of seawater.

In freshwater fish, whose bodies have a higher concentration of salt than the waters around them, the process occurs in reverse. Freshwater fish constantly absorb water into their bodies, so they have no need to drink it; instead, they discharge the excess fluid by copious urination.

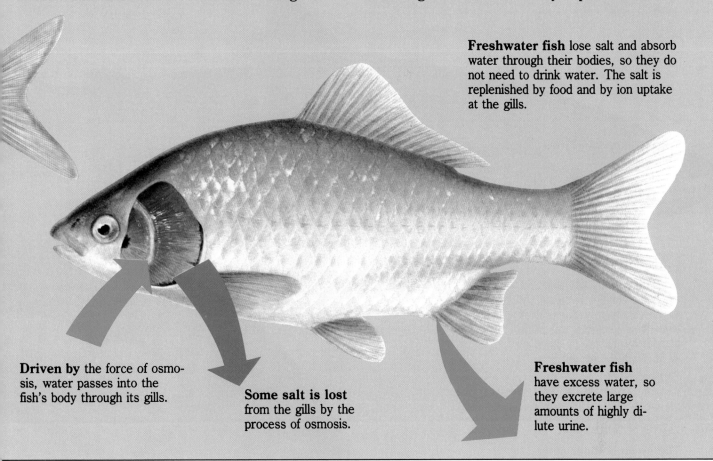

Freshwater fish lose salt and absorb water through their bodies, so they do not need to drink water. The salt is replenished by food and by ion uptake at the gills.

Driven by the force of osmosis, water passes into the fish's body through its gills.

Some salt is lost from the gills by the process of osmosis.

Freshwater fish have excess water, so they excrete large amounts of highly dilute urine.

Quick-change artists

Several species of fish are diadromous—that is, they can exist in both salt water and fresh water. They do this by regulating their body fluids to suit their environment. They drink water—or refrain from it—according to the saltiness of their surroundings. In addition, their gills and kidneys can quickly shift from processing fresh water to salt water, and back again. Salmon, which travel from the ocean to fresh water for spawning, and sturgeons, shads, and lampreys, which live near river mouths, are among these adaptable fish. Some other diadromous species are shown at right.

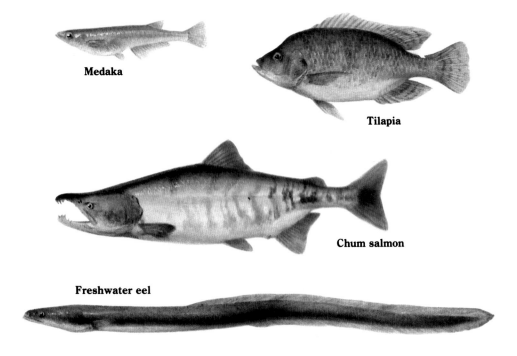

Medaka

Tilapia

Chum salmon

Freshwater eel

11

What Is a Swim Bladder?

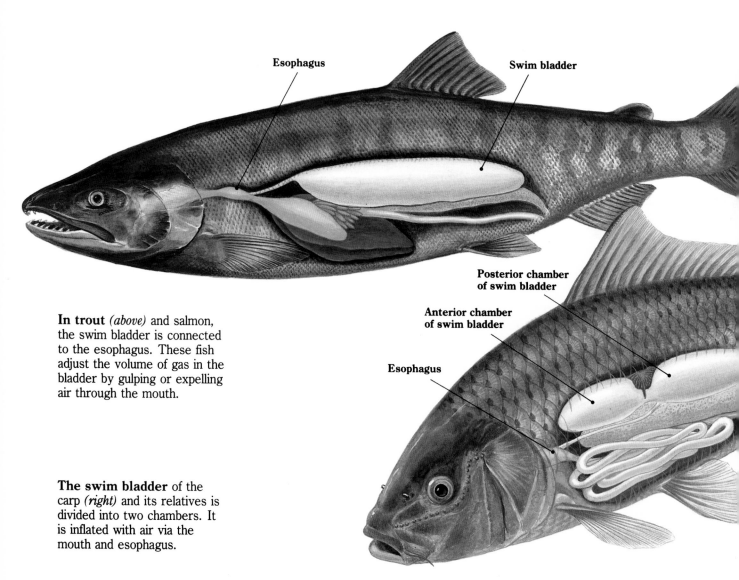

Esophagus

Swim bladder

Posterior chamber
of swim bladder

Anterior chamber
of swim bladder

Esophagus

In trout *(above)* and salmon, the swim bladder is connected to the esophagus. These fish adjust the volume of gas in the bladder by gulping or expelling air through the mouth.

The swim bladder of the carp *(right)* and its relatives is divided into two chambers. It is inflated with air via the mouth and esophagus.

Learning how to float

Newly hatched salmon *(right)* and trout have no gas in their swim bladders, so they cannot control where they float. Because they are easy prey at this stage, the young fish hide among the stones of the river-bed. Later on, they take air into their mouths. In this manner, the fish gradually inflate their swim bladders and regulate their buoyancy.

Baby salmon float to the surface to gulp air, which passes into their swim bladders.

A silvery school of young salmon dart about, gaining buoyancy control.

Inside many fish is a balloonlike organ, known as the swim bladder, or gas bladder, that enables the creature to float effortlessly in the water. Inflated with atmospheric gases such as oxygen and nitrogen, the swim bladder lowers the density of the fish's body to equal that of the surrounding water. As a result, the fish need not swim constantly to avoid sinking. To maintain this neutral buoyancy at various depths, many fish are able to adjust the amount of gas in their swim bladders.

Some fish, though, would be poorly served by neutral buoyancy. Flounder, for example, must be heavy enough to stay on the ocean floor, and some species of tuna dive and ascend too rapidly to allow for gas adjustment. Many bottom dwellers and speedy swimmers therefore have greatly reduced swim bladders—or none at all.

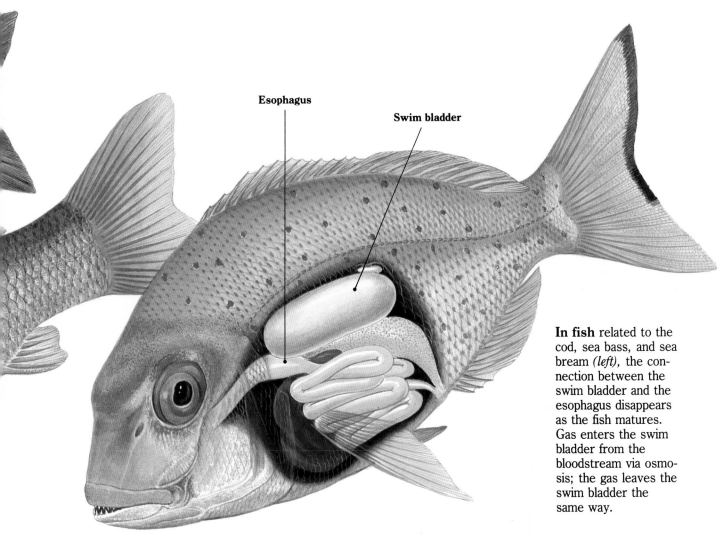

Esophagus

Swim bladder

In fish related to the cod, sea bass, and sea bream *(left),* the connection between the swim bladder and the esophagus disappears as the fish matures. Gas enters the swim bladder from the bloodstream via osmosis; the gas leaves the swim bladder the same way.

A flotation device

When a fish's swim bladder is connected to its esophagus by a narrow tube known as the pneumatic duct, the amount of gas in the bladder is regulated by air through the mouth. In fish whose swim bladder is not attached to the esophagus, the bloodstream helps carry out the gas exchange *(right).* As the blood pressure changes at various depths, the red gland and oval body raise and lower the bladder's gas content.

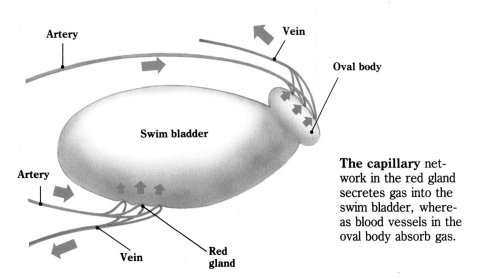

Artery

Vein

Oval body

Swim bladder

Artery

Vein

Red gland

The capillary network in the red gland secretes gas into the swim bladder, whereas blood vessels in the oval body absorb gas.

Are All Fish Cold-Blooded?

Fish are often called "cold-blooded" because most of them are unable to conserve the body heat they generate. They lose most of the heat through their gills; as a result, their body temperature matches the

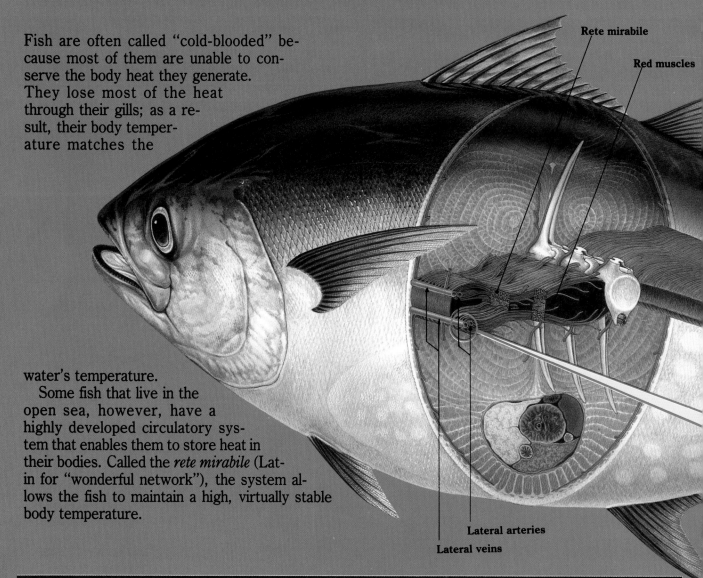

Rete mirabile

Red muscles

Lateral arteries

Lateral veins

water's temperature.

Some fish that live in the open sea, however, have a highly developed circulatory system that enables them to store heat in their bodies. Called the *rete mirabile* (Latin for "wonderful network"), the system allows the fish to maintain a high, virtually stable body temperature.

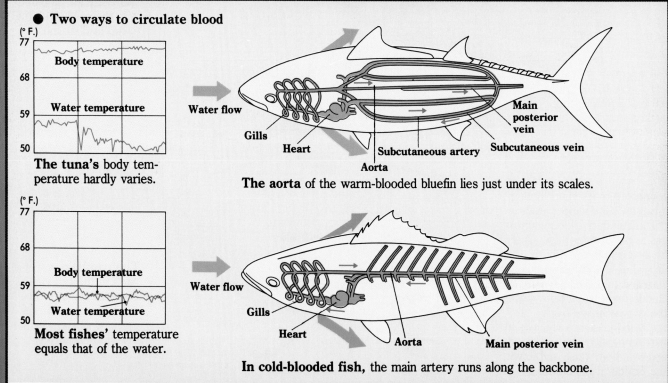

● **Two ways to circulate blood**

(° F.)

77

68

59

50

Body temperature

Water temperature

The tuna's body temperature hardly varies.

Water flow

Gills

Heart

Aorta

Subcutaneous artery

Subcutaneous vein

Main posterior vein

The aorta of the warm-blooded bluefin lies just under its scales.

(° F.)

77

68

59

50

Body temperature

Water temperature

Most fishes' temperature equals that of the water.

Water flow

Gills

Heart

Aorta

Main posterior vein

In cold-blooded fish, the main artery runs along the backbone.

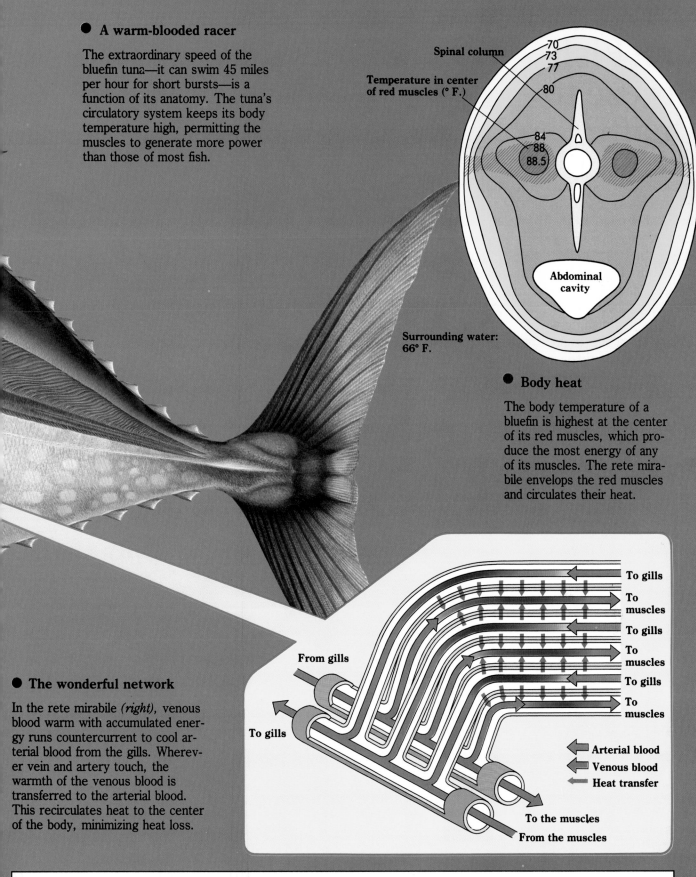

● A warm-blooded racer

The extraordinary speed of the bluefin tuna—it can swim 45 miles per hour for short bursts—is a function of its anatomy. The tuna's circulatory system keeps its body temperature high, permitting the muscles to generate more power than those of most fish.

Spinal column

Temperature in center of red muscles (° F.)

70
73
77

80

84
88
88.5

Abdominal cavity

Surrounding water: 66° F.

● Body heat

The body temperature of a bluefin is highest at the center of its red muscles, which produce the most energy of any of its muscles. The rete mirabile envelops the red muscles and circulates their heat.

To gills
To muscles
To gills
To muscles
To gills
To muscles

From gills

To gills

⇐ Arterial blood
⇐ Venous blood
← Heat transfer

To the muscles
From the muscles

● The wonderful network

In the rete mirabile *(right)*, venous blood warm with accumulated energy runs countercurrent to cool arterial blood from the gills. Wherever vein and artery touch, the warmth of the venous blood is transferred to the arterial blood. This recirculates heat to the center of the body, minimizing heat loss.

Warm-blooded relatives

Most members of the tuna and mackerel families are distinguished by the rete mirabile, the blood network that preserves body heat. Like the fish at right, species with this system are pelagic—that is, they live in the open seas.

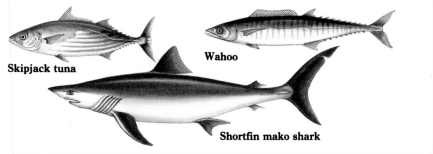

Skipjack tuna

Wahoo

Shortfin mako shark

Can Fish Freeze to Death?

Because the body temperature of most fish mirrors that of their environment, they can freeze to death in icy waters. The body fluids of an ordinary fish will solidify if the surrounding water drops below 31° F. External ice then penetrates the gills and skin, accelerating the fatal freezing process.

One type of fish, however, has adapted to survive in the frigid Antarctic Ocean, where temperatures range from 28 to only 29.5° F. The blood of the notothens contains a natural antifreeze, called glycoprotein, that keeps the fish's body fluids liquid until its body temperature falls below 28° F. The glycoprotein lowers the freezing point of body fluids by blocking the formation of ice crystals.

Far below a ceiling of pack ice, bottom-dwelling spotted notothens *(right)* thrive in the subfreezing waters of the Antarctic Ocean.

Freezing points

The chart at right explains why notothens can withstand water temperatures that most other fish—represented here by the surfperch—cannot. The body fluids of both species contain salt, which lowers their freezing point below that of ordinary water—32° F. But glycoprotein in the blood of the notothen enables it to endure temperatures as low as 28° F.

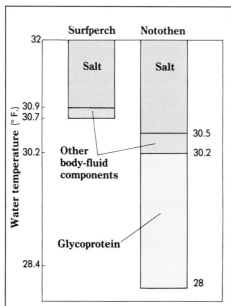

Cold-loving species

Pictured at right are five of the 90 to 100 varieties of notothens that inhabit the Antarctic Ocean. Most notothens are bottom dwellers, but at least two species—including the Antarctic toothfish—have abandoned the seabed to swim and hunt in deep waters.

The icy seas of northern Canada are home to species of cod, bullhead, and flatfish that resist subfreezing temperatures. Like the notothens, these species produce glycoprotein, which lowers the freezing point of their body fluids.

16

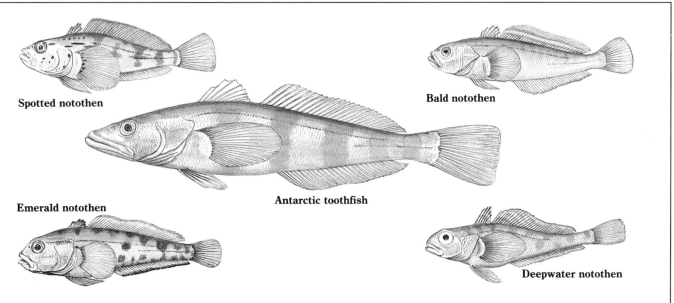

Spotted notothen

Bald notothen

Antarctic toothfish

Emerald notothen

Deepwater notothen

How Do Different Animals Swim?

Of all the creatures that make the water their home, each has its own special way of swimming. Some animals use their entire bodies to generate thrust—the force for forward motion—while others rely mainly on specific parts of the body such as the tail or fins. Depending on its body shape and method of propulsion, an animal may be best suited to accelerating, maneuvering, or high-speed cruising. Or it may manage all three actions fairly well but excel at none. An aquatic creature's mode of swimming, determined by its physical form, dictates how it snares prey and eludes enemies.

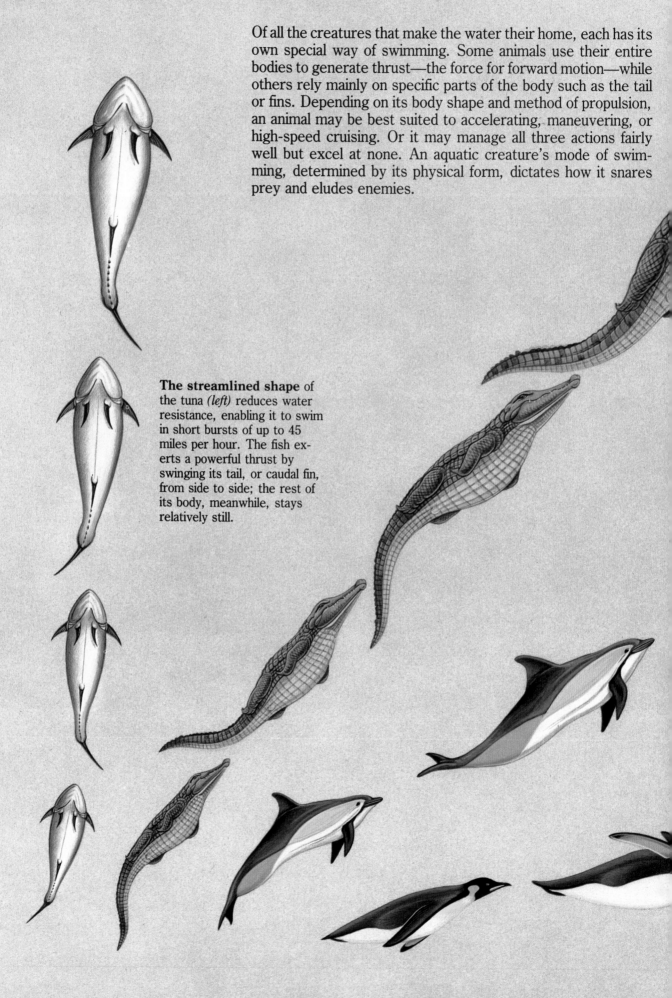

The streamlined shape of the tuna *(left)* reduces water resistance, enabling it to swim in short bursts of up to 45 miles per hour. The fish exerts a powerful thrust by swinging its tail, or caudal fin, from side to side; the rest of its body, meanwhile, stays relatively still.

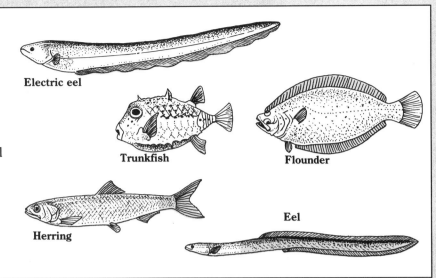

A swim-style sampler

The electric eel swims by waving its anal fin. Other eels slither along by twisting their entire bodies in S-shaped curves. Flounder skim the sea bottom by rippling their dorsal and anal fins, while the trunkfish swings its caudal fin like a pendulum. Herring—like most other fish—throw their bodies into undulating waves in order to swim.

Electric eel

Trunkfish

Flounder

Herring

Eel

Whipping its long, thick tail from side to side, the crocodile cruises through swampy waters powerfully. To reduce drag, the crocodile tucks all four legs into its body as it swims. Its muscular tail allows the croc to attain bursts of speed to catch prey.

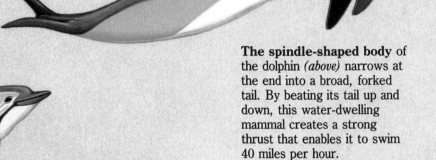

The spindle-shaped body of the dolphin *(above)* narrows at the end into a broad, forked tail. By beating its tail up and down, this water-dwelling mammal creates a strong thrust that enables it to swim 40 miles per hour.

In contrast to their clumsy gait on land, penguins are remarkably agile in the water. The flightless birds flap their flippers the same way ordinary birds move their wings to fly.

19

Why Can Swordfish Swim So Fast?

With its streamlined form and powerful, crescent-shaped tail fin, the swordfish can race through the water at 55 miles per hour. The sweptback tail, characteristic of most fast-swimming pelagic species, is connected to the body by a narrow joint called the caudal peduncle. This enables the swordfish to whip its tail from side to side while keeping the front of its body fairly still—an efficient design that maximizes thrust and minimizes drag. The swordfish's other fins help it steer and keep its balance.

Built for speed, the swordfish *(above)* has a muscular anatomy that allows it to slice through the water with minimal resistance. The body is oval throughout its length.

Judging a fish by its fins

Fin design, which varies enormously from one species to the next *(right),* can reveal how a fish swims or hunts for prey. A tail fin that is rounded or only slightly forked, for example, generates quick acceleration; it often belongs to species that lie in wait to strike at their prey. A crescent-shaped tail fin is the signature of fast-cruising fish, which roam the seas in search of food.

With a swish of its fan-shaped tail, the grouper can dart from its resting place near the ocean floor to nab a tasty morsel that happens by—or to flee from danger.

The tail of the porcupinefish provides propulsion. It also acts as a rudder for turning and lends the fish stability—a crucial survival feature for the rocky waters it inhabits.

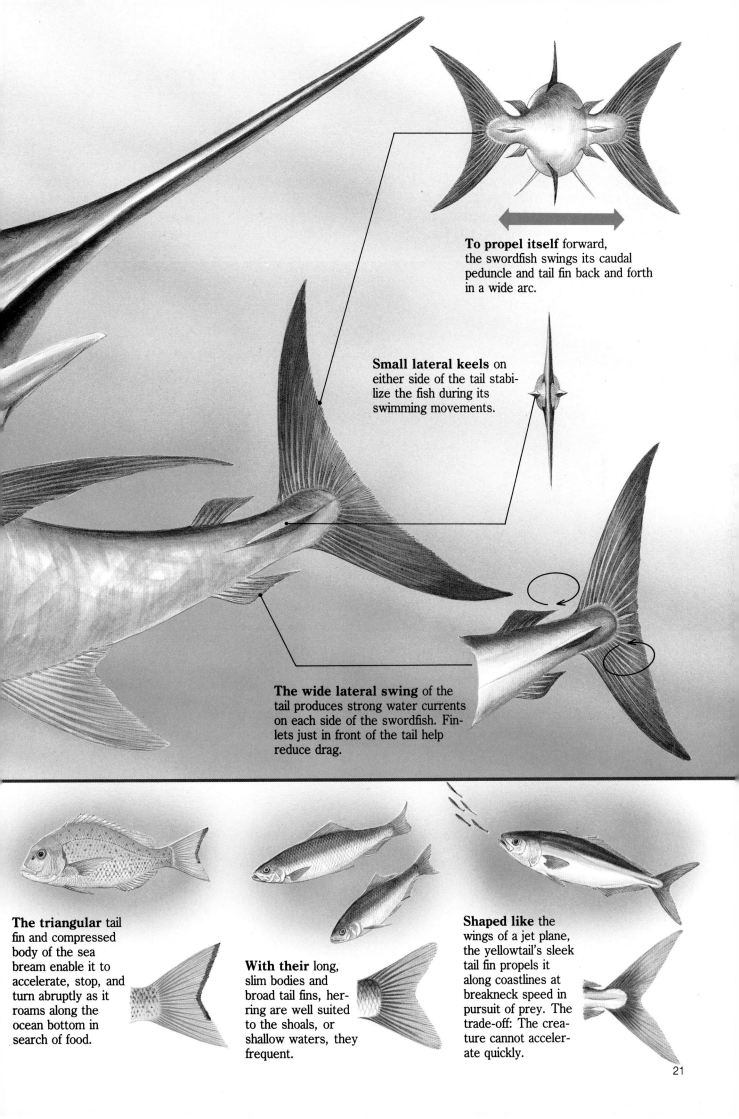

To propel itself forward, the swordfish swings its caudal peduncle and tail fin back and forth in a wide arc.

Small lateral keels on either side of the tail stabilize the fish during its swimming movements.

The wide lateral swing of the tail produces strong water currents on each side of the swordfish. Finlets just in front of the tail help reduce drag.

The triangular tail fin and compressed body of the sea bream enable it to accelerate, stop, and turn abruptly as it roams along the ocean bottom in search of food.

With their long, slim bodies and broad tail fins, herring are well suited to the shoals, or shallow waters, they frequent.

Shaped like the wings of a jet plane, the yellowtail's sleek tail fin propels it along coastlines at breakneck speed in pursuit of prey. The trade-off: The creature cannot accelerate quickly.

How Do Fish Survive Ocean Pressures?

Deep-sea fish lead pressure-filled lives. At 3,000 feet below the surface, the water pressure is 100 times that of sea-level air. In the deepest ocean—some 36,000 feet down—each cubic inch of water is under nearly 8 tons of pressure.

Such pressures would crush a land animal, yet marine creatures are unaffected by them. This is because their body tissues are filled with liquids and dissolved gases that are just as compressed as the surrounding seawater. The internal fluids of deep-sea fish therefore push outward as forcefully as the ocean presses in. Fish with gas-filled swim bladders are especially susceptible to injury from rapid changes in pressure.

Diving apparatus

Many fish that move through zones of varying pressure have a gas-filled swim bladder to aid them in buoyancy control. As a fish swims downward, the pressure of the surrounding water rises and the gases in the bladder are compressed; this decreases the fish's buoyancy, helping it sink. As the fish swims upward, the gases in the bladder expand; this increases the volume of the fish, aiding its rise. The fish must not ascend too quickly; sudden bladder expansion can rupture its insides.

Grace under pressure

A gas bladder gives surface fish a convenient means of diving and rising, but it would be impractical for creatures of the deep: Below 3,000 feet, the gases in a swim bladder reach such high pressures that a steel tank would be needed to contain them at sea level. For this reason, some deepwater species have evolved swim bladders filled with liquid fat, a substance that provides buoyancy but no gas-compression problems. Other fish—the sedentary, bottom-dwelling gulper, for example—have lost their swim bladders through eons of disuse. Creatures like the goblin shark, meanwhile, never developed swim bladders at all. Instead, they have oil-filled livers and lightweight, cartilaginous skeletons that help them adjust to pressure variations and keep them from sinking. A watery, boneless physique is characteristic of many deep-sea dwellers, including some sharks, bighead smoothheads, squid, and anglerfish.

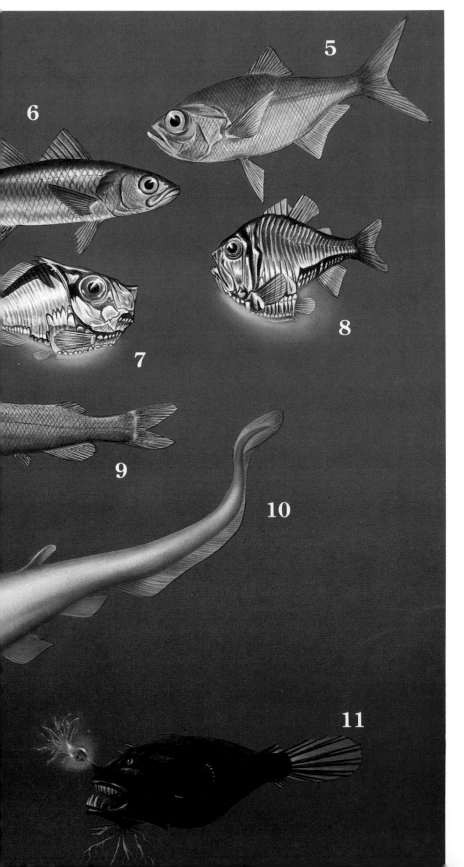

1. During the day, the lanternfish dives as deep as 2,600 feet. At night, however, it skims the surface of the sea, where its luminescence makes it resemble a floating light.

2. Named for its gaping mouth, the gulper reaches depths greater than 3,000 feet.

3. The slickhead is a smooth-skinned fish that migrates between 3,200 and 5,500 feet.

4. The absence of a swim bladder allows many whalefishes to live at depths of 6,000 feet and deeper.

5. An advanced swim bladder lets the alfonsino roam between 300 and 3,000 feet.

6. Japanese bluefish start life in shallow water but dive to 2,000 feet as adults.

7. Thin as a coin, the spiny hatchetfish dives as deep as 1,500 feet.

8. The oblique hatchetfish lives in the twilight realm between 150 and 1,300 feet.

9. Because they lack swim bladders, bighead smoothheads thrive at 1,600 to 4,300 feet.

10. The goblin shark gets its buoyancy from an oil-filled liver, not a swim bladder, as it prowls the Pacific's deep waters.

11. Fierce but slow, the devil angler uses its luminous lure to attract prey in the inky depths between 1,500 and 15,000 feet.

Why Do Some Fish Glow?

In the dimly lit waters 1,000 feet under the surface, the ocean glitters with luminescent creatures. Roughly 96 percent of the creatures that live in this realm emit red, green, blue, violet, or white light from organs known as photophores. These tiny beacons signal danger, lure prey, or call mates. The photophores of species such as the lanternfish are self-illuminating—that is, they create light by secreting a light-emitting substance called luciferin. Other species, like the slender sweeper, get luciferin from the luminous crustaceans they eat. Still others, like the lantern-eye, glow in the dark because luminous bacteria live in their photophores.

1 The "belly light" of the lanternbelly shines because billions of microscopic bacteria live inside it, feasting on the fish's blood. As they feed, the bacteria give off a bright glow.

2 The slender sweeper is luminous thanks to a diet of glowing crustaceans—among them the tiny, blue-glimmering seed shrimp *(far right)*.

3 Chemicals in the photophore of the devilfish produce light that is transmitted through the surrounding water. The creature grabs and eats fish that follow the light to its source.

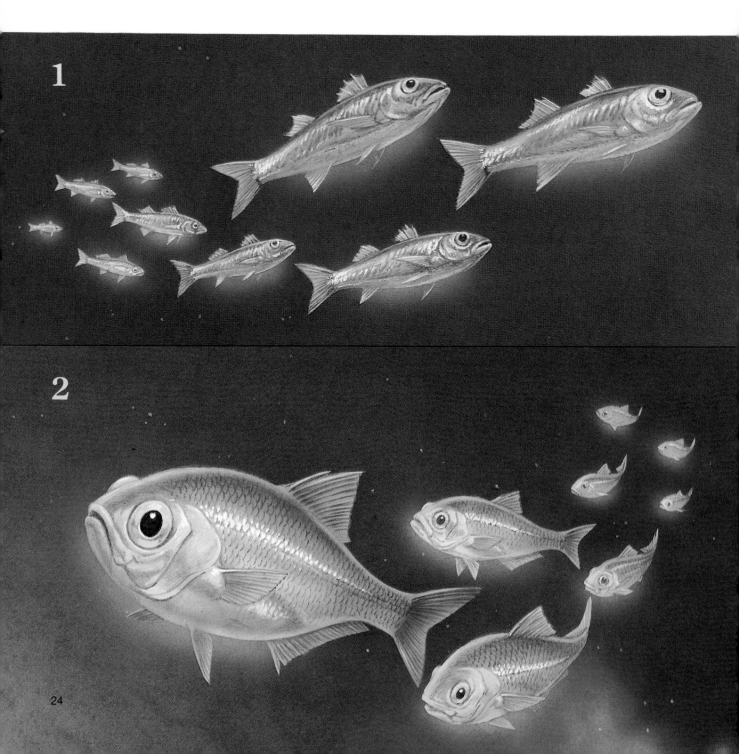

1

2

Tiny vampires

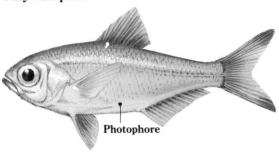

The single photophore of the lanternbelly teems with glowing bacteria that enter through the anus and live on small amounts of the fish's blood.

Photophore

Eating light

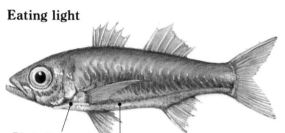

Photophore   Photophore

The slender sweeper has two photophores—luminous organs— which collect a light-emitting chemical from crustaceans the fish has eaten and digested.

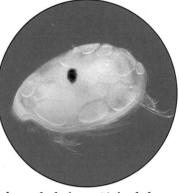

A seed shrimp, ⅛ inch long

3

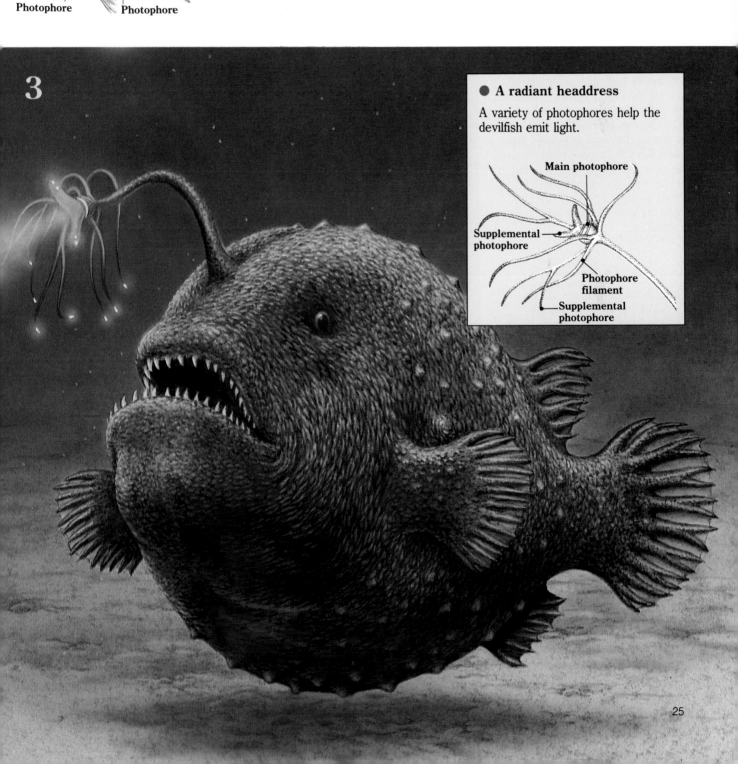

● **A radiant headdress**

A variety of photophores help the devilfish emit light.

Main photophore

Supplemental photophore

Photophore filament

Supplemental photophore

Why Do Luminous Fish Glimmer?

To communicate with other fish and to locate food, many luminous fish flash their body lights on and off. Self-illuminating species like the devilfish blink their lights with nerve signals that start and stop light-producing chemical reactions. But species whose luminescence comes from steadily glowing bacteria must use mechanical means to flicker. The *Photoblepharon* hides its light behind a fold of skin, while the *Anomalops* rolls its photophore inward. The pine cone fish has the simplest blinking technique of all: It hides its photophores under its top lip.

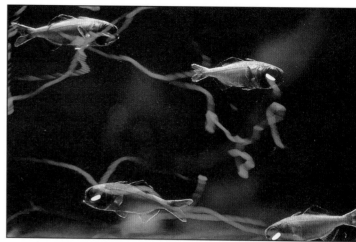

In this time exposure, a school of *Anomalops* create streaks of light that resemble seaweed.

For the pine cone fish *(right)*, blinking its luminous organs is as easy as closing its mouth. When the fish folds its top lip down over the two photophores below its mouth, the light is hidden. But when the fish opens its mouth, the pale green light shines forth. In this way, the steady light produced by the bacteria in the photophores winks on and off, attracting prey.

The *Photoblepharon*—Greek for "eyelid of light"—uses its winking light organ *(opposite page)* to attract prey. Following suit, fishermen in Indonesia's Banda Islands attach the *Photoblepharon*'s photophores to their nets to serve as glowing lures. The severed light organs shine for hours.

Indeed, so bright is the gleam from the *Photoblepharon*'s light organs that skin divers can read their watches by it. The fish itself, however, is unaffected by the glare: Behind each photophore is a black lining that keeps the light out of the fish's eye. The tiny swimmers use their light organs as searchlights, scanning the water around them for food.

Now you see it, now you don't

The *Photoblepharon* emits an eerie blue-green light from a bacteria-filled photophore just below each eye. The fish hides its light by drawing a black eyelidlike membrane up over the photophore *(far right)*. In a relaxed state, the 3-inch-long fish blinks two or three times a minute.

A close relative of the *Photoblepharon,* the *Anomalops (right)* uses specialized muscles to rotate its light organ inward so its glow cannot be seen. At rest, the *Anomalops* turns its photophore inward for five seconds and outward for 10. Victims are attracted to the pulsating light.

When uncovered, the light organ gleams.

Raising the membrane dims the light.

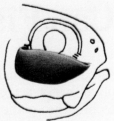
Covered up, the photophore looks dark.

Turned outward, the organ shines.

Muscles roll the photophore inward.

Fully rotated, the organ looks black.

Can Fish Breathe out of Water?

Although most fish need a steady supply of water across their gills in order to breathe, some species are able to survive out of water for as long as several hours. One such creature, found mainly in the mangrove swamps of Southeast Asia, is the mudskipper *(right)*. Not only can it breathe air, but also it can skip across muddy shores—hence its name—and even shin its way up tree trunks. Out of water, the 6-inch-long, bug-eyed mudskipper absorbs life-sustaining oxygen through its skin and gill chambers. This process is described below.

Looking less like a fish than a frog, a mudskipper suns itself on a log. Mudskippers leave the water for hours on end to eat crustaceans and other shellfish stranded by low tide.

The anatomy of a mudskipper

The mudskipper is well suited to life on land. Its skin can absorb oxygen from the air. Water stored in special gill chambers—shown at right in a bottom view of the fish—keeps the gills moist so that they, too, extract oxygen from the atmosphere. Underwater, the mudskipper's gills function in the usual manner.

The mudskipper uses its thick, leglike pectoral fins to propel itself across land. Its pelvic fin comes in handy for gripping such surfaces as rocks and trees, while its protruding, swiveling eyes give it a sweeping view of food and enemies.

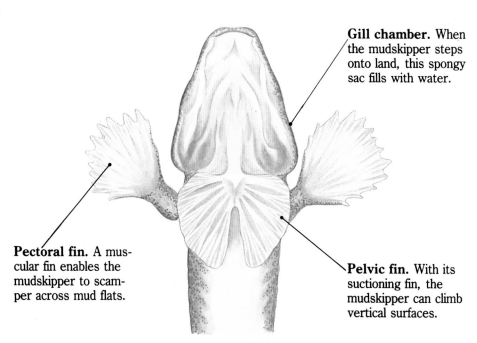

Gill chamber. When the mudskipper steps onto land, this spongy sac fills with water.

Pectoral fin. A muscular fin enables the mudskipper to scamper across mud flats.

Pelvic fin. With its suctioning fin, the mudskipper can climb vertical surfaces.

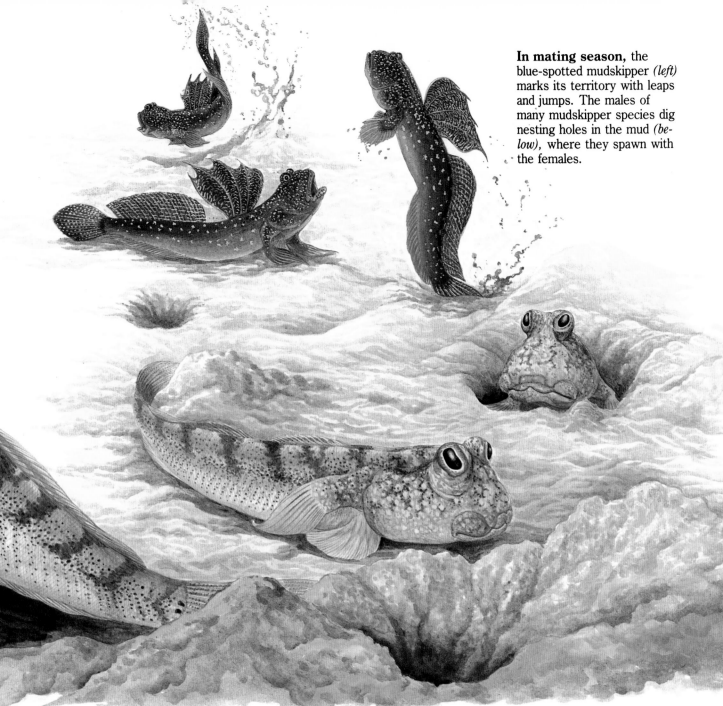

In mating season, the blue-spotted mudskipper *(left)* marks its territory with leaps and jumps. The males of many mudskipper species dig nesting holes in the mud *(below)*, where they spawn with the females.

A land-loving perch

Like the mudskipper, the climbing perch is right at home on land. The most striking feature of this freshwater fish is its upper gill respirator—a convoluted respiratory organ that lies in the gill chamber and enables the creature to breathe air.

Because the climbing perch inhabits stagnant, oxygen-poor bodies of water, it must supplement gill respiration with air breathing in order to survive. It spends much of its time crawling across land from one pool to the next, flexing its powerful tail against the ground and pulling itself along with its pectoral fins.

Air passes through the mouth and over the upper gill respirators *(right)*, which absorb oxygen into the bloodstream.

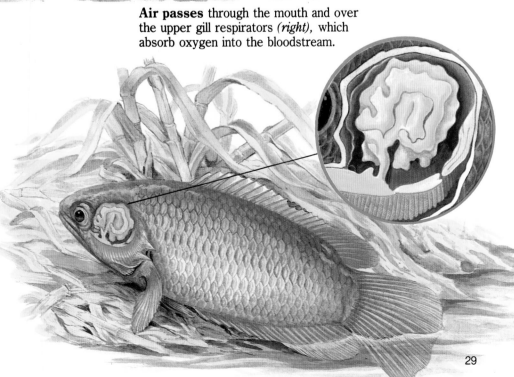

29

How Do Fish Survive When Rivers Dry Up?

Every year, many rivers and lakes in the tropics dry up for three to four months or more. Amazingly, not all of the aquatic life in them dies. The lungfish, for example, has thrived in such desiccated environments for over 300 million years.

At the onset of the dry season, the air-breathing creature burrows into the mud *(below)*, covers itself with a mucous cocoon *(right),* and enters a dormant state known as estivation. Tunnels to the surface supply the lungfish with air.

A look at the lungfish

The lungfish differs from other fish in several basic ways. First, it is equipped with lungs, not gills; the lungs enable it to draw oxygen from the atmosphere. Second, it has a simplified heart designed to increase the efficiency of its lungs. Third, a pair of limblike fins help the lungfish drag itself over dry land.

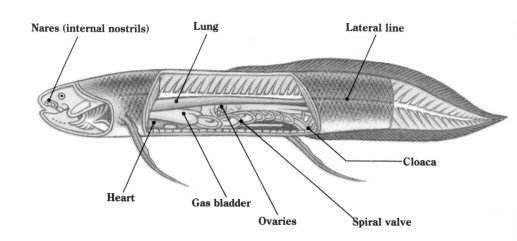

Nares (internal nostrils)   Lung   Lateral line

Cloaca

Heart   Gas bladder   Ovaries   Spiral valve

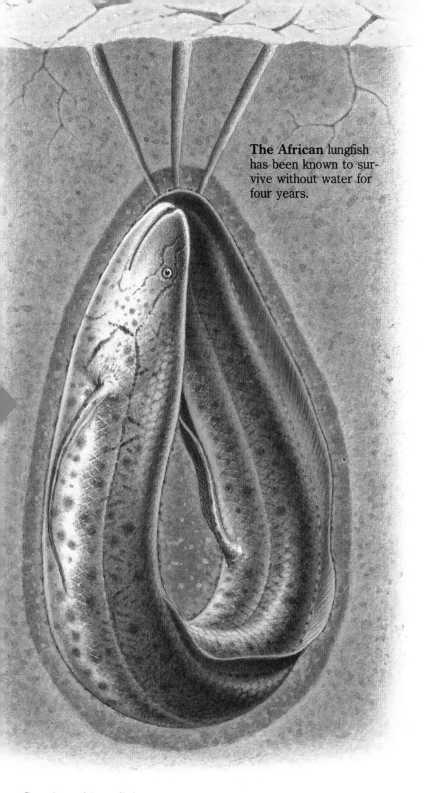

The **African** lungfish has been known to survive without water for four years.

Species of lungfish

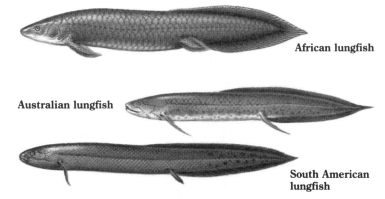

African lungfish

Australian lungfish

South American lungfish

The South American and African lungfish can withstand long droughts, but the Australian species cannot survive out of water.

Hatching a plan for survival

Some fish, such as the medakas shown below, ensure the survival of their species by laying eggs that withstand drought. Just weeks after the eggs are laid, the dry season begins and the adult fish perish. But as water levels rise at the beginning of the rainy season, the eggs hatch. The young fish then grow to maturity in time to lay eggs of their own, and the cycle starts anew.

As the dry season approaches, a spawning pair of medakas *(above)* deposit their fertilized eggs in the sand or silt of a riverbed.

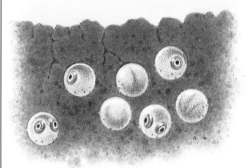

Secure at a depth where the sand retains life-giving moisture, the medaka eggs grow gradually. Not long before hatching, they enter a resting phase that conserves water.

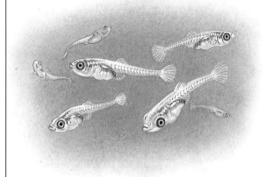

Young medakas emerge from their eggs at the start of the rainy season. Before their brief lives end, they will breed to guarantee a new generation of fish.

How Do Sperm Whales Dive So Deep?

Whales are air-breathing mammals, so they spend most of their time near the surface of the ocean. To find squid, plankton, and other food, however, they dive far beneath the waves. The adult male sperm whale dives the deepest of all, plunging as far down as three-quarters of a mile and staying submerged for up to two hours on a single breath.

The sperm whale performs these feats thanks to an immense reservoir of oil in his head. Called spermaceti oil, the fluid enables the whale to adjust his buoyancy at varying depths, much as a fish uses its swim bladder to ascend and descend. Because spermaceti oil changes density—and buoyancy—with temperature, the whale cools the oil with seawater when he wants to dive; to rise again, he warms the oil with his blood.

Powered by his flipperlike tail, a bull sperm whale dives toward the seafloor at 400 feet per minute. He can ascend at a rate of almost 500 feet per minute.

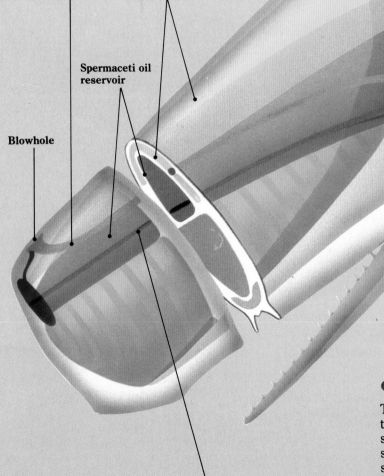

Left nasal passage

Nasal muscle

Spermaceti oil reservoir

Blowhole

Nasopalatine groove

Right nasal passage

● A body built for diving

The bull sperm whale is a master conservationist—of oxygen, that is. While the whale is submerged, his heartbeat slows dramatically, saving oxygen. Half of the whale's oxygen supply is stored in a substance called myoglobin in the muscles. Blood cells distribute the remaining oxygen to vital organs. Together these strategies enable the sperm whale to use 90 percent of the oxygen he inhales; humans, by contrast, use only 20 percent.

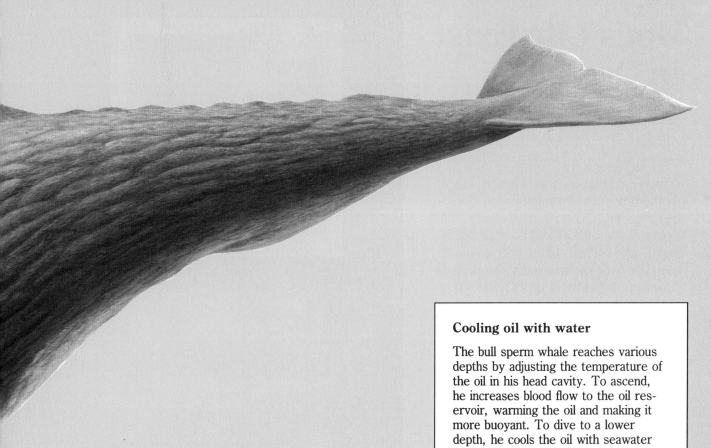

For the love of squid

The sperm whale's diving habits help satisfy his hunger for giant squid. A bull whale may lurk in the ocean depths for up to two hours until a squid floats by. When that happens, the whale nabs it in his toothy jaws and swallows it whole. Examinations of the stomachs of sperm whales have turned up 35-foot-long squids weighing 40 pounds. A typical bull eats as much as a ton of food in a day; small wonder, then, that he can grow to weigh 45 tons.

Cooling oil with water

The bull sperm whale reaches various depths by adjusting the temperature of the oil in his head cavity. To ascend, he increases blood flow to the oil reservoir, warming the oil and making it more buoyant. To dive to a lower depth, he cools the oil with seawater as shown below.

a

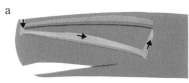

b

Flexing its nasal muscles, the whale draws cool seawater *(blue)* into its right nasal passage (a). It expels the water by contracting the same passage (b).

c

d

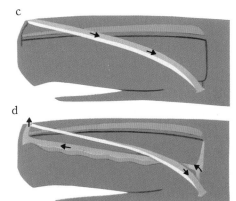

Flaring its left nasal passage, the whale sucks water into the nasopalatine groove (c). The water floods the right canal, where nerve tissue expels it (d).

Does Life Exist in Ocean Trenches?

Snaking along the floor of Earth's seas is a ribbon of canyonlike trenches. At the bottom of this lightless and near-freezing abyss 4 to 7 miles deep, scientists have discovered a variety of bizarre life forms: blind crabs; giant, blood-red clams; and hairy, mouthless tubeworms that may grow to be a dozen feet tall.

These creatures owe their existence to hot-water vents—also called black smokers—that dot the trenches. The vents spew out black clouds of hot, sulfur-laden seawater, heating the surrounding ocean from its normal 34° F. to 55° F. This attracts sulfur-eating bacteria, which multiply at astounding rates—more than four million occupy a teaspoon of seawater—and become food for higher organisms.

1. The mundiopsis, a deep-sea relative of the hermit crab, scoots backward.

2. Hemoglobin, the pigment of red blood, colors the flesh of these foot-long clams.

3. Because they live in perpetual dark, these crabs have no need for eyes.

4. Each of these blood-filled tubeworms absorbs food through 300,000 tentacles.

Tiny galatheid crabs feast on bacteria around a vent.

Sea anemones

Hornito extinct vent

Black smoker

4

They all live in a bacterial world

Bacteria support the oases of life around hot-water vents. The microbes live on oxygen, carbon dioxide, and hydrogen sulfide in the water gushing up through the vents. Multiplying inside the vents, the bacteria accumulate in mats until currents wash them into the open, where they are eaten. With the conch, however, they live in harmony *(right)*.

Until the discovery of trench colonies, scientists believed sunlight is the energy source for all life. They now know that heated water and sulfuric compounds can play the same role.

Sulfur-eating bacteria in the gills of a conch produce organic matter that keeps the mollusk alive.

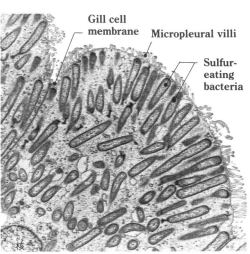

Gill cell membrane

Micropleural villi

Sulfur-eating bacteria

Bacteria fill the gill cells of a conch.

Why Do Crabs Blow Bubbles?

Like fish, crabs breathe with their gills. Water flows into each gill chamber through an opening near the base of the crab's chelipeds, or pincers. The gills extract oxygen from the water, then expel the water through siphons—two small holes in the roof of the mouth.

Almost all crabs that live in the ocean breathe by circulating water in this way. Some crabs, however, have adapted to living on dry land; to conserve precious moisture, they breathe by recycling water through their gills. Water expelled from the siphons streams down the body, picking up fresh oxygen, then flows back through the gills. If the water supply is not replenished, the recycled water becomes sticky, making bubbles when it emerges from the siphons.

Using gills on dry land

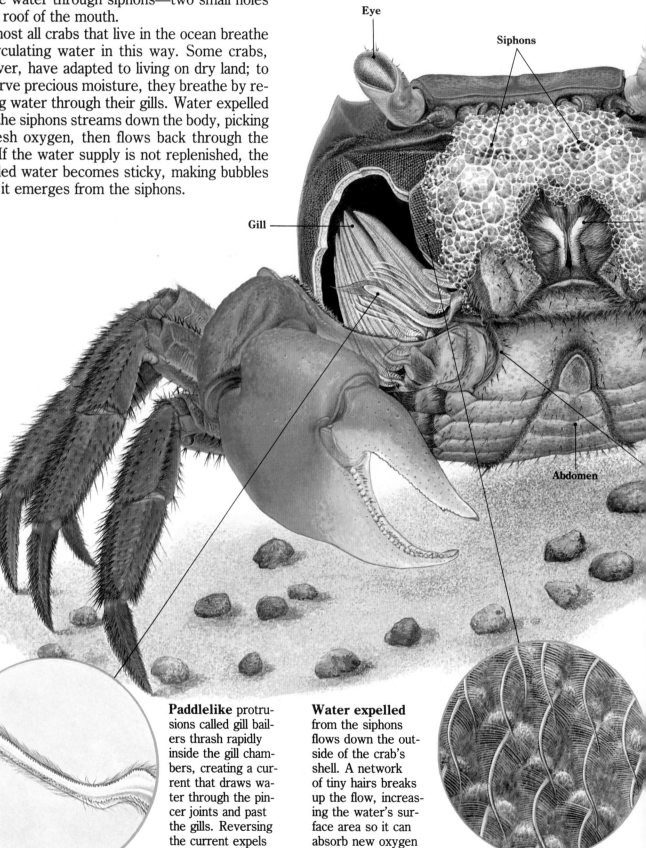

Eye

Siphons

Gill

Abdomen

Paddlelike protrusions called gill bailers thrash rapidly inside the gill chambers, creating a current that draws water through the pincer joints and past the gills. Reversing the current expels sand from the gills.

Water expelled from the siphons flows down the outside of the crab's shell. A network of tiny hairs breaks up the flow, increasing the water's surface area so it can absorb new oxygen from the air.

Once the gills have strained out the oxygen, the water is ejected through two siphons in the roof of the crab's mouth. Water spurts in jets from a nearly submerged crab *(left);* if the crab is on dry land, the water may form bubbles as it emerges.

A land crab *(above)* blows bubbles from its siphons. Many species of crabs live in dry areas, some of them far from the sea. The crabs replenish the water in their gills with rain and dew, returning to the ocean only for spawning.

Cheliped (pincer)

Mouth

Bubbles for every occasion

Bubbles coming from the siphons of a crab indicate that the animal is having trouble breathing. When water is abundant, the crab seldom blows bubbles. Situations such as the ones shown below, however, may cause bubbles to appear.

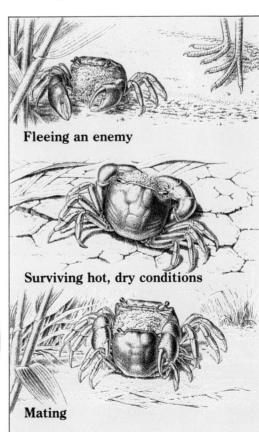

Fleeing an enemy

Surviving hot, dry conditions

Mating

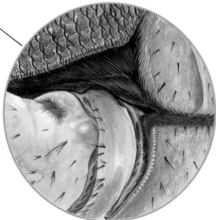

After flowing down the crab's sides, freshly oxygenated water enters the gill chambers through the chela joint openings. The inlets are covered by tiny hairs that filter out debris.

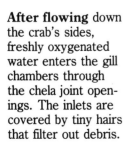

Which Crabs Invade Other Shells?

Parasites in search of a palace, pea crabs regularly inhabit the shells of living bivalves such as clams and mussels. The hosts they most often choose are the short-necked clam *(below)*, the blue mussel, and the common clam. Pea crabs have also been found living in the bodies of shrimps, sand dollars, sea squirts, and sea cucumbers.

Another bivalve invader is the oyster crab. The adult female oyster crab measures about half an inch across; the male is less than half that big. As soon as a female emerges from the larval stage and takes on the shape of an adult, she enters a host shell. Where the young adult males spend most of their time is unknown; they enter the host shells only during the breeding season, to seek out the females within.

Once inside a host shell, the female crab grows until she is too large to leave. Her hard shell, no longer needed for protection, becomes soft and nearly transparent. The female crab may live in the host for several breeding seasons. Then, after the crab dies, her decomposing body is flushed harmlessly out of the shell.

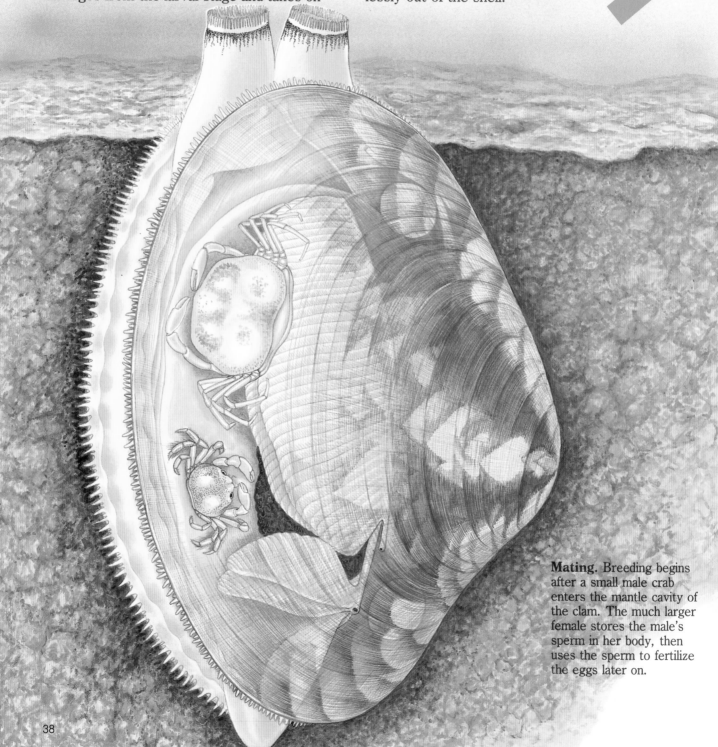

Mating. Breeding begins after a small male crab enters the mantle cavity of the clam. The much larger female stores the male's sperm in her body, then uses the sperm to fertilize the eggs later on.

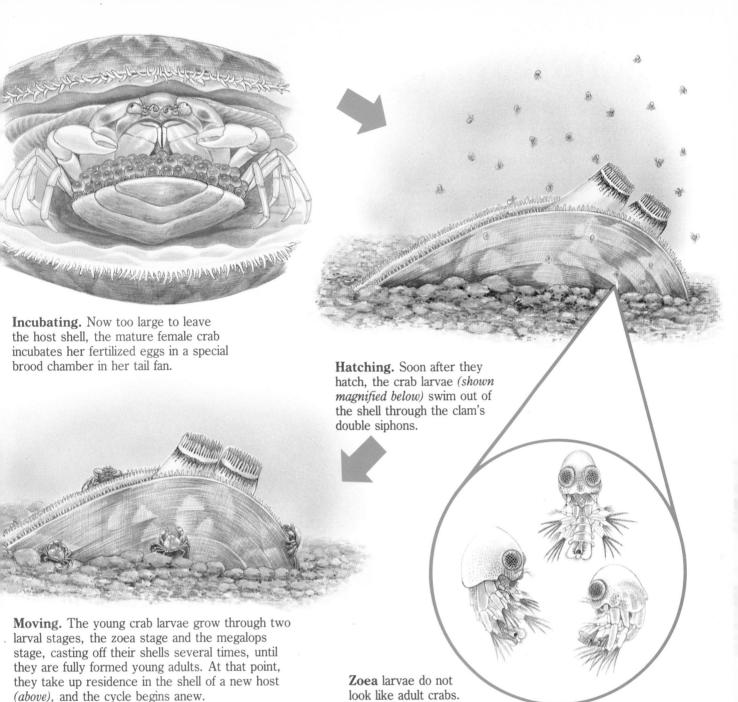

Incubating. Now too large to leave the host shell, the mature female crab incubates her fertilized eggs in a special brood chamber in her tail fan.

Hatching. Soon after they hatch, the crab larvae *(shown magnified below)* swim out of the shell through the clam's double siphons.

Moving. The young crab larvae grow through two larval stages, the zoea stage and the megalops stage, casting off their shells several times, until they are fully formed young adults. At that point, they take up residence in the shell of a new host *(above)*, and the cycle begins anew.

Zoea larvae do not look like adult crabs.

How to live with a crab

Many crab species live outside their host rather than inside it. For protection, some nestle among the stinging filaments, called nematocysts, of sea creatures such as anemones *(right)* and corals. Such living arrangements do not always benefit both parties. Crabs that attach themselves to sea urchins, for example, may hamper their hosts' movements and cause their spines to break off.

Other crabs, however, actively help their hosts. Crabs that reside among coral chase away predators; they also clean unwanted debris off the coral. This relationship, in which two dissimilar organisms prosper by living in close contact, is known as symbiosis.

Crab living in a sea anemone

Crab parasite on a sea urchin

How Do Barnacles Cling to Rocks?

Barnacles are a varied breed—almost 900 species have been counted so far—yet they have one trait in common: They cling tenaciously to the same spot throughout their adult lives, be it a rock, a piece of driftwood, a strand of kelp, or even a large marine mammal. Some barnacles fasten themselves to the bottoms of ships; in this stowaway fashion, they have spread around the world.

The soft body parts of a barnacle are enclosed in a hard shell shaped like a volcano. The shell is made of small, overlapping wall plates, supported by a flat base. Every barnacle begins life as a free-swimming larva. It then settles on a more or less flat surface and attaches itself permanently by secreting a gluey substance from cement glands at the base of its first pair of antennae. This cement is the strongest in the world; it has been studied for possible use as a dental adhesive.

Barnacles are hermaphroditic—that is, each individual has both male and female reproductive organs. Thus a barnacle can fertilize itself. Procreation usually occurs, however, when the male organ of one barnacle enters the receptive female organ of a nearby barnacle, which then produces eggs.

Searching. Each cypris larva drifts through the ocean, feeding on plankton until it reaches a suitable site to settle.

Touching down. Finding a surface where other barnacles live, the larva places its first pair of antennae on the rock.

Settling. At its chosen site, the barnacle larva secretes cement to attach itself, then molts to become a young adult.

A barnacle's hold on life

Barnacles on rocks in the intertidal zone are exposed to air at low tide. To conserve water during this interval, a barnacle closes its operculum. When the tide comes in, the barnacle opens and extends its cirri—six pairs of forked appendages with fine hairs that comb edibles from the water. After a while, the barnacle retracts the cirri, carrying the food back to its mouth.

Once a barnacle is attached, it remains in place for the rest of its life. The outer shell grows at a slow, steady pace.

High and dry on seaside rocks, barnacles await the return of the tide. Various species live at different heights; the barnacles that are best adapted for exposure to air occupy the highest rocks.

Covered with seawater by the tide, barnacles open their mouths and extend their cirri. The fine hairs on the cirri collect microscopic, nutritious plankton from the water.

Spawning. When the eggs hatch, swarms of nauplius larvae swim out of the parent's shell and into the open water.

Molting. The nauplius larvae *(above)* molt several times to become cypris larvae *(left)*.

Penis

Mating. One barnacle extends its male sex organ to enter a neighbor's female organ. The fertilized barnacle stores its eggs in its mantle cavity, protecting them until they hatch.

Barnacle build

A barnacle's shell, which protects the softer parts of its body, consists of four to eight wall plates that slope inward toward a hole at the top. Though usually conical, the shell may grow tall and narrow in tight quarters. Two hard plates make up the operculum; muscles move these plates to open and close the aperture. The barnacle's soft body is similar to that of a shrimp, with the abdomen uppermost. Food-gathering cirri are connected to the abdomen.

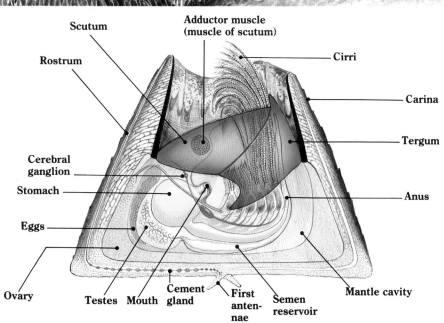

Scutum

Rostrum

Adductor muscle (muscle of scutum)

Cirri

Carina

Tergum

Cerebral ganglion

Stomach

Anus

Eggs

Ovary

Testes Mouth Cement gland

First antennae

Semen reservoir

Mantle cavity

Can Jellyfish Swim?

The oceans of the world teem with 200 varieties of jellyfish. Some of the creatures are shaped like disks. Others are shaped like combs. And still others trail long, transparent tentacles beneath them in the water.

Many jellyfish are excellent swimmers. They swim by repeatedly contracting and relaxing the muscles beneath their umbrella-shaped bells *(right);* each contraction shoots water out of the bell, propelling the jellyfish in the opposite direction. Much of the time, however, jellyfish simply float in the water, hanging just below the surface and drifting with the currents.

In a disk-shaped jellyfish, the underside of the bell has a mouth and a number of long oral arms. The bell itself is made of a jellylike material, mesoglea, which is held in place by an inner and outer body wall. Around the rim of the bell are tentacles armed with minute organs called nematocysts; when stimulated by prey or predator, the nematocysts shoot out stinging filaments.

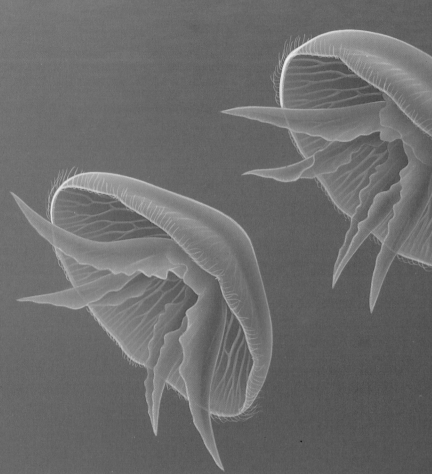

Different strokes for different folks

Some jellyfish use specialized structures to get around. The Portuguese man-of-war, for example, has a gas-filled float that acts as a sail. The sail measures up to a foot long. When it catches the wind, the sail carries the man-of-war across the water at an angle of about 45 degrees to the wind direction.

Other species can deflate their floats to sink below the surface during storms. Still others have a specialized swimming bell behind the float. Expanding and contracting in rhythm, the bell takes in water and then shoots it out through tubes, propelling the jellyfish like a jet.

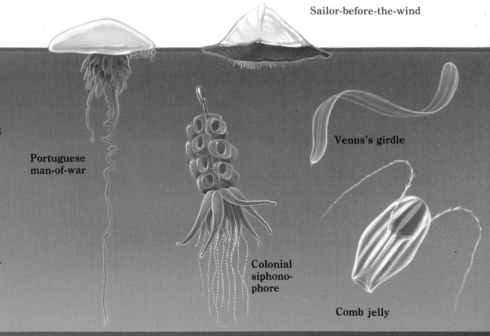

Sailor-before-the-wind

Portuguese man-of-war

Colonial siphonophore

Venus's girdle

Comb jelly

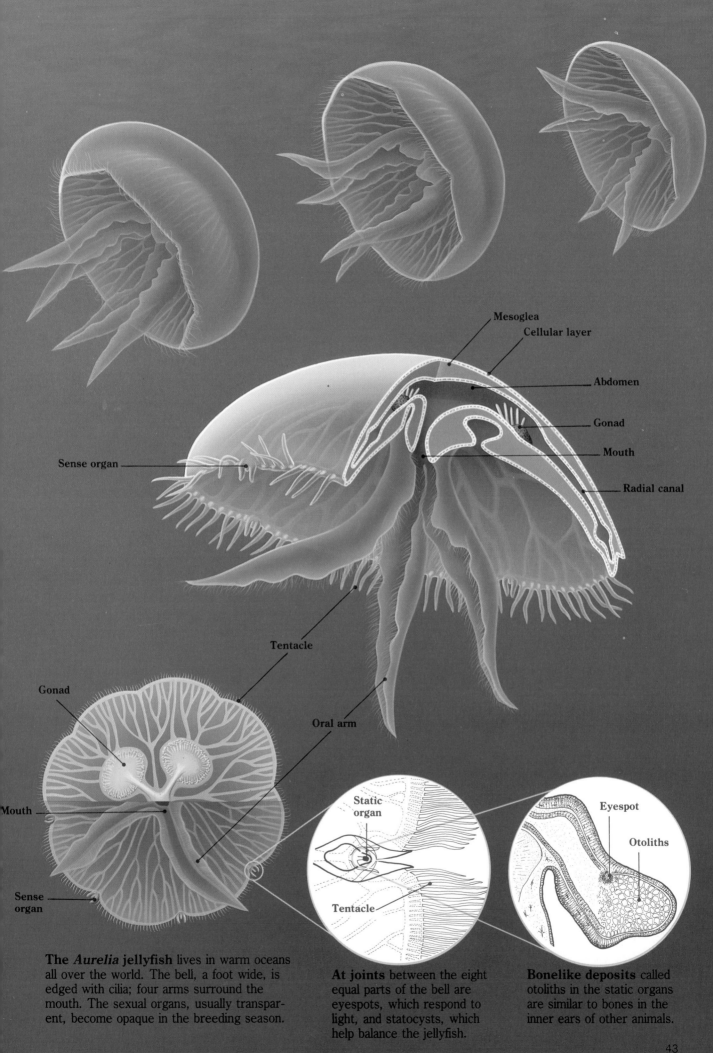

Mesoglea

Cellular layer

Abdomen

Gonad

Mouth

Radial canal

Sense organ

Tentacle

Oral arm

Gonad

Mouth

Sense organ

Static organ

Tentacle

Eyespot

Otoliths

The *Aurelia* jellyfish lives in warm oceans all over the world. The bell, a foot wide, is edged with cilia; four arms surround the mouth. The sexual organs, usually transparent, become opaque in the breeding season.

At joints between the eight equal parts of the bell are eyespots, which respond to light, and statocysts, which help balance the jellyfish.

Bonelike deposits called otoliths in the static organs are similar to bones in the inner ears of other animals.

2

Sensible Swimmers

Like the creatures that live on land, fish need a wide array of senses to find food and avoid predators. The five traditional senses—sight, smell, taste, touch, and hearing—are all present in fish, but the way they are employed often differs dramatically from their use by land animals.

Vision is crucial to many species of fish. Because fish live underwater, the lenses in their eyes are spherical, a shape that can focus light rays in the thicker medium of water. The fish-eye lens provides a wide field of vision. Smell, too, is essential to most fish, especially in locat-

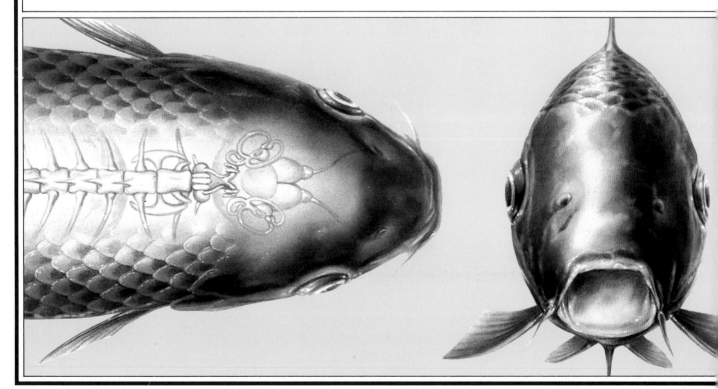

ing prey; it can be effective through more than half a mile of water. Taste, by contrast, is a close-range sense. It is used to distinguish between edible and poisonous food.

The sense of touch in fish often surpasses that in land animals. Above and beyond simply feeling the objects they come in contact with, fish can sense vibrations in the water. Thus the thrashing of an injured fish can be both felt and heard by other fish nearby.

The more advanced aquatic creatures boast sensory organs that verge on the exotic. Sharks, for example, use electrical receptors not only to navigate but also to seek out prey. And river dolphins emit clicks, whistles, and chirps whose echoes guide them through muddy waters.

Pictured above is a gallery of diverse eye styles; from left to right, they represent the octopus, wrasse, requiem shark, four-eyed fish, sea bream, and ray. Shown below are the sensory organs a carp uses to hear (left) and smell (right).

How Does the Sense of Smell Work?

Fish rely on their sense of smell to locate food, evade enemies, navigate through river systems, and find sexual partners. Their sense of smell is a chemical system triggered by odors—that is, by substances dissolved in water. When an odor enters the fish's nostrils, or nares, it stimulates sensory cells lining the surface of olfactory plates there. In response, these cells send signals to the olfactory center in the brain, which identifies the odor. Amino acids and other organic substances excite two senses, smell and taste, thus providing even more detailed information to the fish.

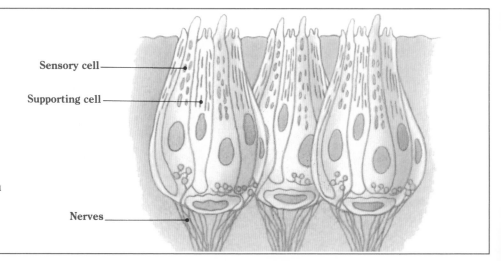

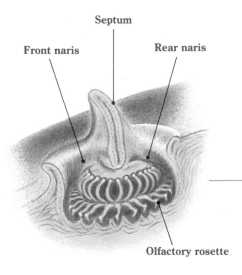

Septum

Front naris

Rear naris

Olfactory rosette

The olfactory system of the carp shown at right is typical of most bony fishes. On either side of the carp's snout is a pouch lined with nerve tissue that is able to detect odors. Water enters the pouch through the front naris and exits through the rear naris.

A fish's sense of taste

Fish have taste buds covering the tongue, the roof of the mouth, the throat, and—in some species—the lips and outer-body surfaces. Like humans, fish can distinguish sweet, sour, salty, and bitter tastes. They also have nerve endings that can sense amino acids and fatty acids; detection of these substances, given off mainly by shellfish, assists the fish in locating their prey.

Sensory cell

Supporting cell

Nerves

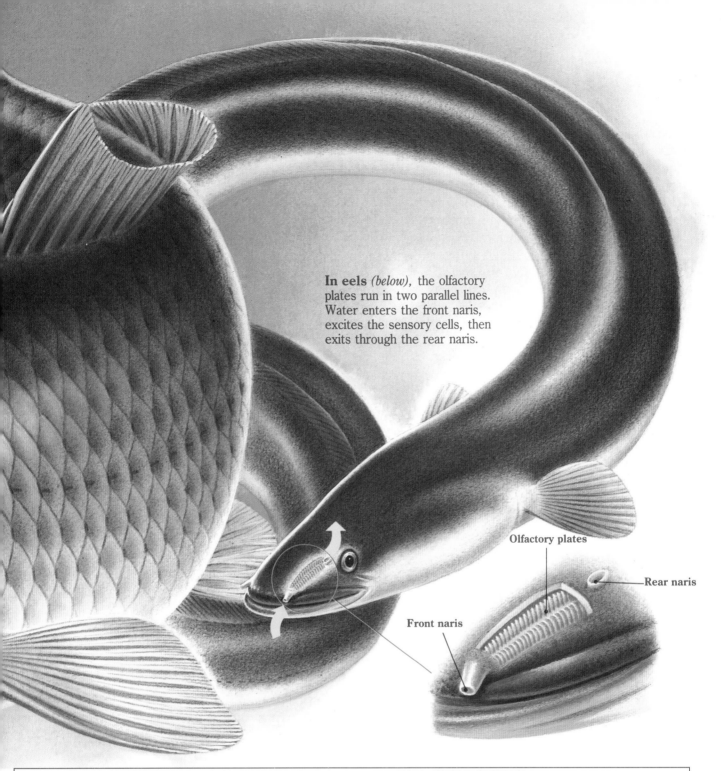

In eels *(below),* the olfactory plates run in two parallel lines. Water enters the front naris, excites the sensory cells, then exits through the rear naris.

Olfactory plates

Rear naris

Front naris

Tasting with fingerlike fins

The unusual fish known as a gurnard *(right)* uses both touch and smell to determine if the creatures it encounters will make a tasty meal. Extending three fingerlike spines from both of its pectoral fins, the fish probes the sand and silt of the river bottom for tiny creatures such as shellfish. Special cells located on the spines identify amino acids and other chemicals from shellfish but don't respond to salty, sweet, or bitter tastes from other creatures.

How Do Fish See?

With their eyes typically set on the sides of their heads, fish can see all around their bodies. This high degree of peripheral vision, as it is called, allows them to flee predators who attack from below or behind. Only in the area directly in front of them, however, do fish have three-dimensional vision. In that field, they can focus on objects at different distances.

Fish can also see color—up to a point. Red and yellow are easily perceived, but green, blue, and black are indistinguishable. Most fish live at a certain depth, so their color vision is tuned to the dominant color in their surroundings.

A fish-eye field of view

Fish have a blind spot that prevents them from seeing objects in this narrow band *(gray)*.

Fish looking forward have a narrow three-dimensional field of view in which they can focus on objects.

Fish see with three-dimensional vision in the area in front of them *(orange)*.

Different ways to see the world

Even though fish live in a radically different environment than humans, their eye structure is somewhat similar. This is because both evolved from the same ancestors—fishes that lived some 500 million years ago.

Fish can see clearly underwater because both the water and the fish's cornea refract light at the same angle. The eyes of many fish feature a spherical lens set in an ellipsoidal eyeball. They can focus on nearby objects in front of them and distant objects to the side.

The human eye cannot focus underwater, where light refracts differently than it does in air. Humans have a biconvex lens that changes thickness to focus on objects far or near.

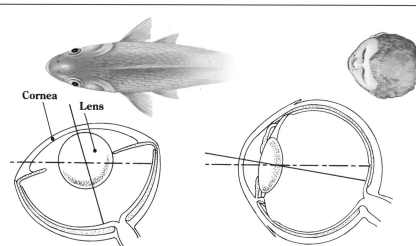

Cornea

Lens

In fish, the visual and optical axes are almost perpendicular. The visual axis *(dashed line, above)* is aligned toward the front, while the optical axis *(solid line)* connects the center of the cornea to the center of the lens.

In humans, the visual and optical axes are almost aligned. This reflects the position of the eyes in a single plane on the front of the face rather than on the sides of the head.

The structure of a fish eye

The eyes of a fish, like those of all vertebrates, are crucial sensory organs that provide a view of both predators and prey. Fish eyes function by focusing light through a lens and onto the retina, the inner lining of the eye. The retina is made up of rods and cones—special photosensitive cells that turn light into nerve impulses and transmit them to the brain, which then forms an image.

Whereas a human eye focuses by changing the thickness of a lens, a fish's eye focuses by moving the lens forward and backward. This movement is performed by a tiny muscle called the *retractor lentis*. When the muscle relaxes, the lens moves closer to the fish's snout, and nearby objects become clear. But when the muscle tightens, the lens is pulled toward the back of the eye, and distant objects come into focus.

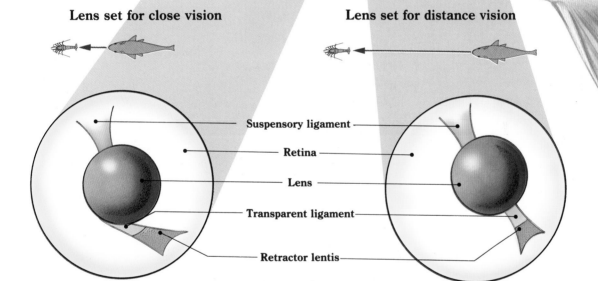

Lens set for close vision

Lens set for distance vision

Suspensory ligament

Retina

Lens

Transparent ligament

Retractor lentis

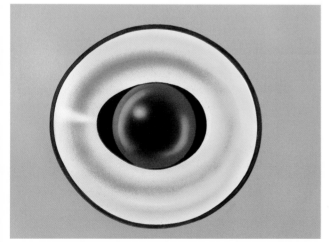

To focus on an object nearby, the fish relaxes its retractor lentis muscle, moving the lens toward the snout.

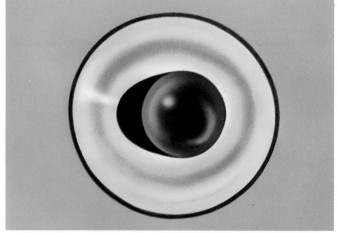

To focus on a distant object, the retractor lentis contracts, pulling the lens back as much as half a millimeter.

What Are Four-Eyed Fish?

Swimming along the surface of freshwater streams and rivers in Central and South America is a strange creature called the four-eyed fish. This 10-inch-long relative of the guppy appears to have a total of four eyes, one stacked upon the other on either side of its head. In reality, the fish has only two eyes, but each eye is divided into an upper and lower half. As the fish swims just beneath the surface, the top half of each eye juts above the water, searching the air for flying insects. The bottom half, meanwhile, remains submerged, scanning the water for meals and predators. This remarkable feat is possible because the lens of each eye is specially adapted for bifocal vision, allowing the four-eyed fish to focus clearly on objects in both the aerial and aqueous worlds.

● Keeping an eye out

Because light refracts differently in air than it does in water, humans cannot see clearly underwater and most fish cannot see well through air. The species shown at right, however, are notable exceptions. Both spend a considerable amount of time out of the water, and have therefore evolved unique lenses that enable them to see through air. Flying fish, for example, have a lens that is flat, like a pane of glass; this permits them to see when they leap from the water. The weedfish, fond of resting atop exposed reefs in the Galápagos Islands, sees through a prism-shaped cornea whose flat surfaces face forward and back.

● Nature's bifocals

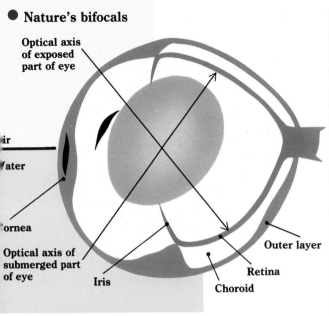

Optical axis
of exposed
part of eye

ir

Water

Cornea

Optical axis of
submerged part
of eye

Iris

Outer layer

Retina

Choroid

ght entering the aerial half of the lens strikes the
ver retina; light from the water hits the upper retina.

As the fish swims near the surface, the upper half of each eye
searches for flying insects; the lower half peers underwater.

Shaped like a three-sided
pyramid with the sides facing
forward, backward, and down,
the cornea of a flying fish
(above) provides clear vision in
those three directions both in
and out of the water.

A prism-shaped cornea lets
the weedfish *(right)* see when
it climbs onto reefs. The cornea is divided vertically, so the
fish can see in front and behind
but not to the side.

Do Fish Have Ears?

Fish ears—impossible to see from the outside because they are contained within the skull—resemble the inner ears of human beings. The ears detect sounds and keep fish balanced and oriented in their three-dimensional underwater world. The ears also help fish sense changes in speed and direction as they swim.

The inner ear of a fish includes three looping canals that are sensitive to changes in pressure. At the base of these canals are the otoliths—hard, pebblelike bones floating in fluid. When the fish moves, the otoliths drift in the fluid, touching sensory hair cells that in turn send electrical impulses to the auditory part of the fish's brain.

Some fish also use their swim bladders *(pages 12-13)* to hear. The bladder senses pressure changes caused by motion in the water; then, through vibrations, it transmits those changes to the ear via a chain of small bones called Weberian ossicles.

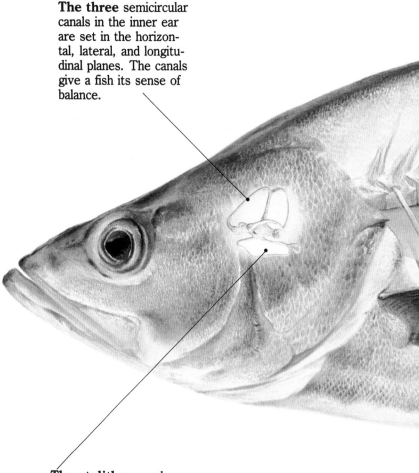

The three semicircular canals in the inner ear are set in the horizontal, lateral, and longitudinal planes. The canals give a fish its sense of balance.

The otoliths—a collection of small bones—signal changes in speed, allowing the fish to hear.

An unusual sound sensor

Minnows, catfish, carp, and some other related fish hear sound through a vibrating bladder of air located just behind the head. This balloonlike organ, called the swim bladder, is connected to the auditory system by a set of small bones known as the Weberian ossicles. When a sound wave moving through the water strikes the fish, it causes the bladder to vibrate. The swim bladder magnifies the vibrations, which are then carried forward by the ossicles to the inner ear.

In fish such as drums, catfish, and others, the swim bladder is a source of sound: By vibrating the tissues that surround the swim bladder, the fish can communicate through the water to other fish nearby.

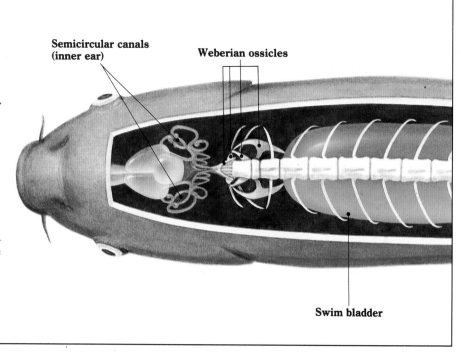

Semicircular canals (inner ear)

Weberian ossicles

Swim bladder

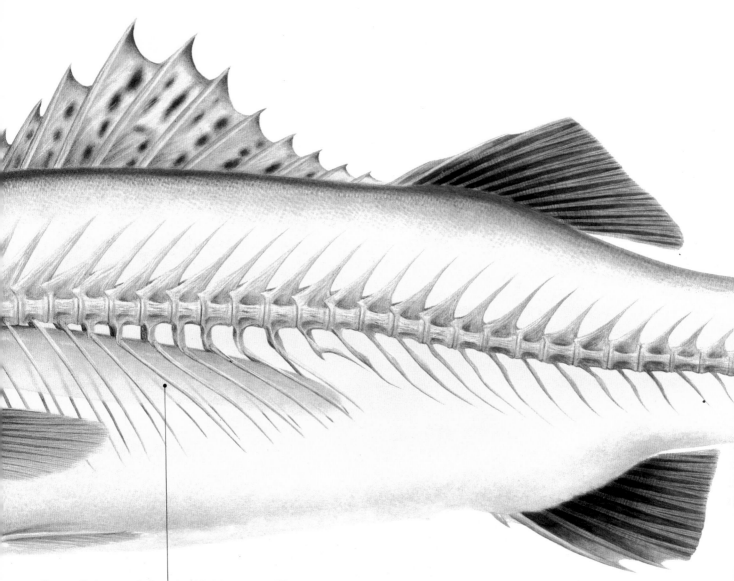

Some fish use their swim bladder to amplify sound. When struck by sound waves, the swim bladder vibrates, magnifies these vibrations, and sends them to the inner ear. Other fish use their swim bladder to produce sound.

Sound and the lateral line

Running nearly the entire length of a fish's body is a third sensory organ, the lateral line, that is a component of the auditory system. A complex system of sensitive hairs embedded in pits or canals, the lateral line can detect even the smallest disturbance in the surrounding water. The lateral line determines the precise direction of the disturbance, warning the fish when it is about to swim into a rock or some other object.

When bent by the motion of water, the hairs protruding from the lateral line send impulses to the fish's brain indicating which way they are bending. These signals prove especially useful in helping the fish find prey.

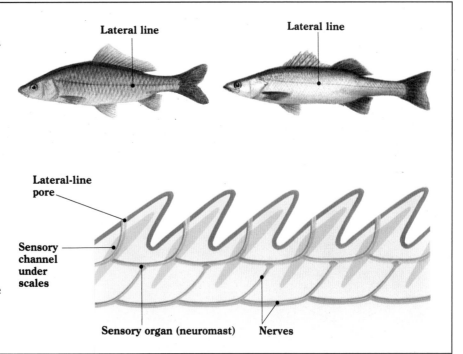

Lateral line

Lateral line

Lateral-line pore

Sensory channel under scales

Sensory organ (neuromast)

Nerves

How Do Sharks Find Their Prey?

Sharks, the most efficient hunters in the sea, use a wide range of senses to track down their prey. Yet sharks are not just eating machines; they can go for days and even weeks without a bite of food. In captivity, they consume only 1 to 2 percent of their body weight each day.

A shark relies on its sense of hearing to first detect a possible prey. The thrashing of an injured fish, for example, can be heard by some sharks more than a quarter mile away. As the shark closes to within 100 yards, its acute sense of smell helps lead it closer to the target. At 100 feet or so, the lateral line—a vibration-sensing organ—helps pinpoint the prey. Not until the shark is 50 feet away, however, can its limited eyesight distinguish the prey. Finally, just inches from its victim, the shark uses an electrical sensing system to close in for the kill.

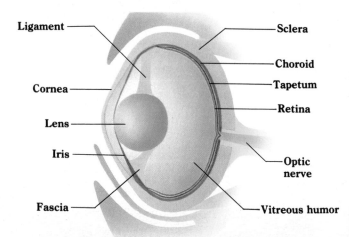

Although sharks cannot make out details very clearly, their light-sensitive eyes see well in dim light. The nictitating membrane can be raised to control the amount of light entering the eyes. They may have color vision.

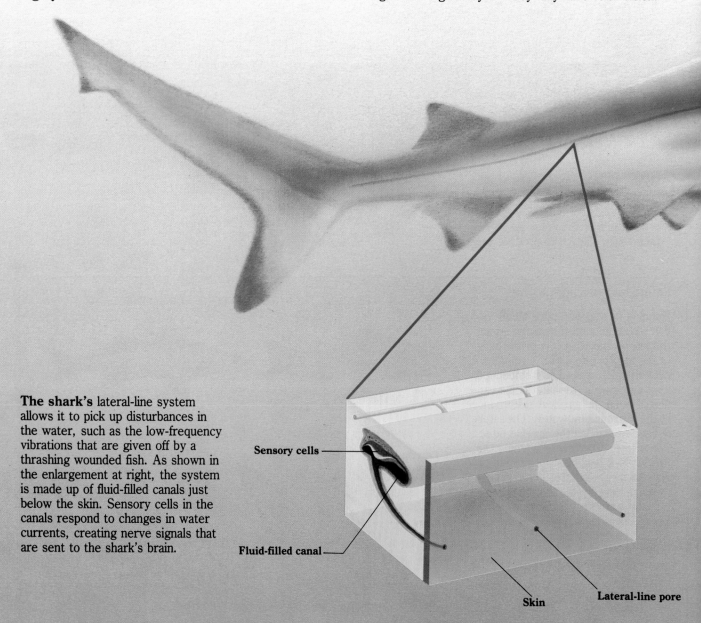

The shark's lateral-line system allows it to pick up disturbances in the water, such as the low-frequency vibrations that are given off by a thrashing wounded fish. As shown in the enlargement at right, the system is made up of fluid-filled canals just below the skin. Sensory cells in the canals respond to changes in water currents, creating nerve signals that are sent to the shark's brain.

Sensory cells

Fluid-filled canal

Skin

Lateral-line pore

Sharks are extremely sensitive to sound. Hair cells in the inner ear convert sound waves to nerve impulses and send them to the brain.

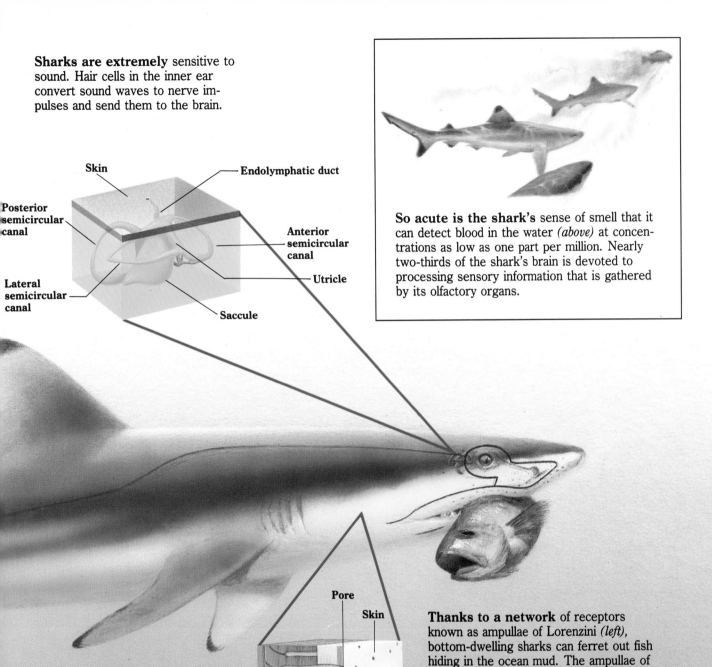

Skin

Endolymphatic duct

Posterior semicircular canal

Anterior semicircular canal

Lateral semicircular canal

Utricle

Saccule

So acute is the shark's sense of smell that it can detect blood in the water *(above)* at concentrations as low as one part per million. Nearly two-thirds of the shark's brain is devoted to processing sensory information that is gathered by its olfactory organs.

Pore

Skin

Nerves

Ampulla of Lorenzini

Thanks to a network of receptors known as ampullae of Lorenzini *(left)*, bottom-dwelling sharks can ferret out fish hiding in the ocean mud. The ampullae of Lorenzini detect the weak electrical field generated by the fish. By sensing the Earth's magnetic field, the ampullae also help the shark navigate.

Blind shark's buff

To test the bioelectric sensing system of a bottom-dwelling shark, researchers buried a living fish under the ocean floor. The shark quickly found the fish. The researchers then buried both a dead fish and a living fish, and covered part of the shark to keep it from using sight, smell, or water vibrations to locate the prey. Using only its electrical receptors, the shark found the living fish but ignored the dead one. When another live fish was wrapped in polyethylene to shield its bioelectric field, the shark could not discover it. Finally, two electrodes that generated an electrical field similar to one from a fish were buried and quickly attacked by the shark.

Even with its other senses blocked, a shark was able to home in on the bioelectric emissions of buried—but living—prey.

The shark failed to pick up the emissions of a living, buried fish wrapped in polyethylene, which masked the fish's electrical field.

What Is an Electric Fish?

Both the electric eel of South America and the electric catfish of Africa stun their prey with a powerful shock generated by a special electric organ. Other fish, too, use electricity to find and capture food, though in less dramatic fashion. They also produce strong electric pulses that help them to detect and deter predators.

The freshwater African fishes known as mormyrids, or elephant fishes, activate gel-filled canals on their bodies to create electrical fields in the surrounding water. When another fish enters this field, it produces distortions that enable the mormyrid to determine the intruder's size and location. Using this information, the mormyrid will follow or flee the newcomer. Inanimate objects in the water also disrupt the electrical field, so the fish can use the field to navigate.

The African electric fish shown below, a relative of the mormyrids, has long, thin electric organs on the rear half of its body. These organs create an electrical field around the fish. The fish's sensitive skin, meanwhile, picks up distortions in this electrical field and senses the fields generated by other fish.

A fishy security system

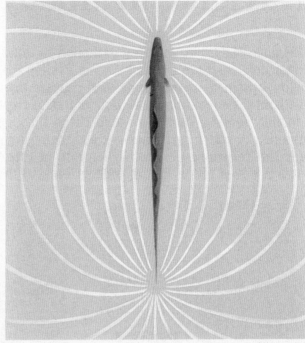

Electric organ

The African electric fish generates an electrical field around its body that is equally strong in all directions. Like an underwater alarm, any distortion of the field alerts the fish to the approach of other creatures.

An electric sense

Objects that have little or no electrical conductivity do not deflect the fish's electrical field. They therefore go unnoticed by the fish.

A fish that is more conductive than the surrounding water distorts the electrical field. Sensing this intruder, the electric fish will move toward or away from it.

Piscatorial stun guns

The electric eel *(right)* and the electric catfish produce enough electrical voltage to knock other fish senseless. An electric eel can build up a charge of around 600 volts—enough to stun even fast-moving prey. Electric catfish have similar power; fishermen who catch them in nets can receive arm-numbing shocks. These painful episodes led to the early belief that God was protecting the catfish caught in their nets.

How Can River Dolphins Navigate in Muddy Water?

Because they live in water darkened by mud, river dolphins have evolved a sonar system that allows them to sense, hunt, and capture prey without ever seeing it. To search for food, a river dolphin sends a series of rapid clicks through the murky water surrounding it and then listens for the returning echoes. When one of these sound waves hits a fish, it reflects back to the dolphin's crescent-shaped upper jaw. Changes in the time-lag and frequency of the echoes lead the dolphin to its quarry.

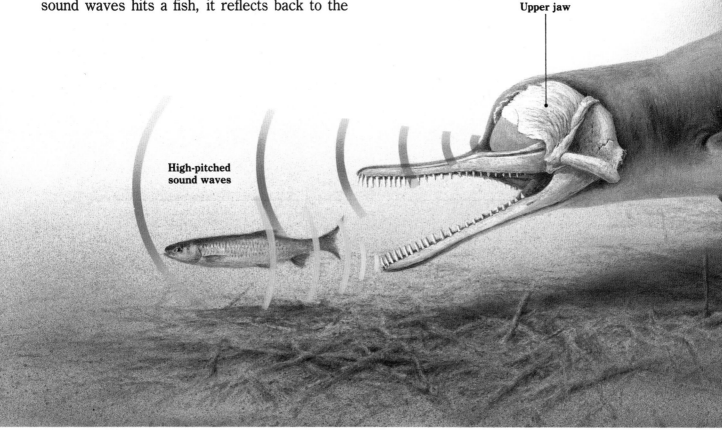

Upper jaw

High-pitched sound waves

Doing the sidestroke

As shown at right, dolphins in the Ganges River of India sometimes turn on their sides to swim. This unusual position allows the nearly blind dolphins to orient themselves: They brush the river bottom with one pectoral fin while sensing surface light through the opposite eye.

In the eons-long course of adapting to its dark environment, the Ganges dolphin lost the crystalline eye lens that would let it focus images. It can, however, distinguish between light and dark.

Species of river dolphins

The five known species of river dolphins live in muddy rivers in Asia and South America. Compared with their saltwater relatives—dolphins, porpoises, and whales—river dolphins have a relatively primitive anatomy.

Because they live near humans, river dolphins are especially vulnerable to disruption by development. They have even been hunted on occasion. Such trends threaten the river dolphin's survival.

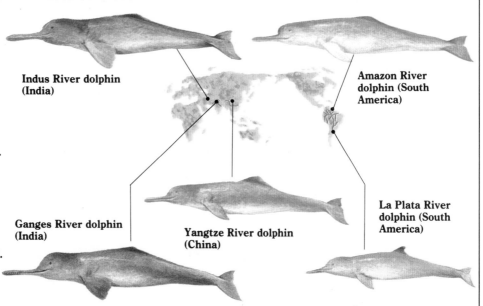

Indus River dolphin (India)

Amazon River dolphin (South America)

Ganges River dolphin (India)

Yangtze River dolphin (China)

La Plata River dolphin (South America)

What Can an Octopus See?

Most marine invertebrates—that is, most sea creatures that lack a backbone—have poor eyesight. The octopus, however, sees its surroundings extremely well. It does not have to rely on its sense of smell, taste, or touch to find crabs and other food.

A dense network of nerves leads from each eyeball to one of two optic lobes beneath the tough outer mantle of the octopus. The lobes attach to the brain of the octopus—a concentrated mass of nerves that forms an unusually elaborate control center for an undersea invertebrate.

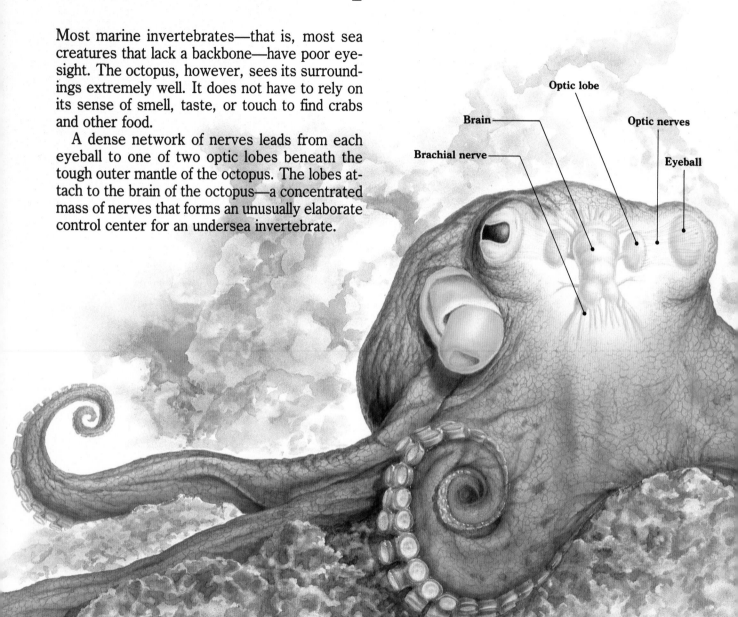

Optic lobe

Brain

Optic nerves

Brachial nerve

Eyeball

Raiding the crab jar

An octopus uses its eyesight, not its sense of smell or touch, to find food. Researchers discovered this by putting a crab—an octopus delicacy—in a glass jar with a cork stopper; though the octopus could see the crab, it could not smell or touch it. The octopus looked at the crab, encircled the jar with some of its eight tentacles, and pulled out the stopper to get the crab. An octopus is also capable of unscrewing a metal lid from a jar to get at a crab inside. Some types of octopus stun crabs and other prey by injecting them with poison from their mouths.

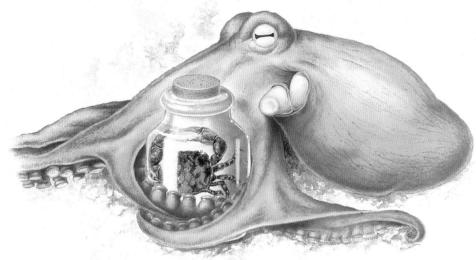

A view to a kill

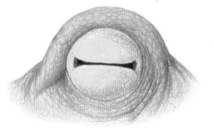

An octopus pupil has the shape of a horizontal rectangle. In bright light, the pupil closes to a slit *(above, left)*. In dim light, it dilates *(above, right)* so the octopus can view its surroundings.

An octopus eye

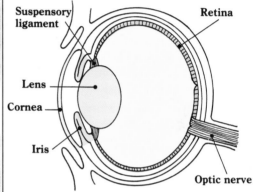

Suspensory ligament
Retina
Lens
Cornea
Optic nerves
Iris

A human eye

Suspensory ligament
Retina
Lens
Cornea
Iris
Optic nerve

So sophisticated is the octopus eye that it resembles the human eye in many ways. The key difference is the optic nerve. In humans, the optic nerve leads from light-gathering rod cells to the brain in a single bundle. In an octopus, each rod has its own optic nerve, resulting in a jumble of nerves that suggests a ball of string. Like many fish, octopuses focus their vision by moving the lens forward or back.

Sharp-eyed and many-armed

An octopus can discern the size and shape of enemies and prey, provided they are not too small or too close. To determine how well an octopus can recognize patterns, scientists showed an octopus two different patterns. When the octopus approached the first pattern, it was given food; when it approached the second, it received a small electric shock. The octopus soon learned to differentiate the two patterns. Later on, it succeeded in distinguishing test objects of differing shade, size, orientation, and shape.

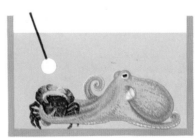

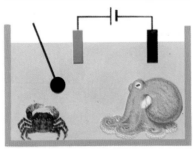

An octopus was shown a white circle; upon moving toward the circle, it received its favorite food.

When the octopus reached for the crab after being shown a black circle, it was given a shock.

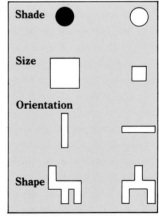

Shade
Size
Orientation
Shape

With training, the octopus could distinguish between the patterns above.

3
Mating, Reproduction, and Growth

Aquatic creatures reproduce in many different ways. Most fishes have a breeding season. During this time, the female leaves masses of small eggs suspended in the water or deposits them in a protected place; the male then scatters sperm over the eggs. The fertilized eggs are often left to fend for themselves. Certain species of fish can even change sex, with males becoming females or vice versa as needed *(pages 68-69)*.

Patterns of spawning—a term used to describe mating that results in fertilization—vary greatly among aquatic animals. Some fish lay a

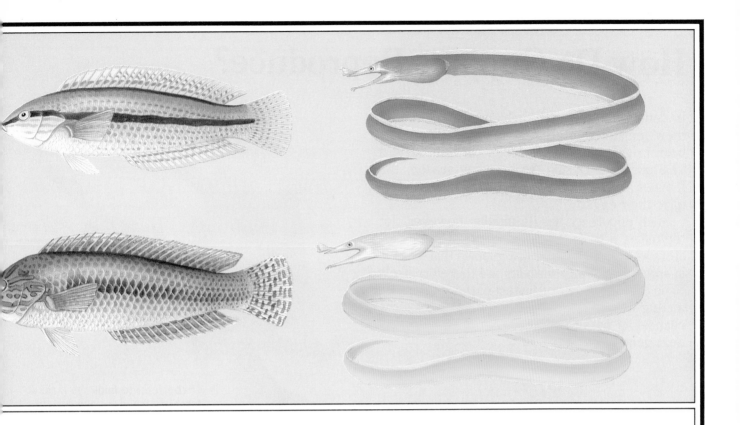

few large eggs, then protect them until they hatch—and occasionally after. Several frogs and toads carry their eggs, and later their young, on their backs or in their abdomens.

The larvae of most marine invertebrates are so small that they can be seen only through a microscope. These tiny larvae spend their young lives floating on the surface while undergoing metamorphoses that transform them into adults; the mature creatures then travel to the place where they will spend the rest of their lives.

Many sea-dwelling mammals leave the ocean during the breeding season to give birth on land. Once their young are grown, the adults return to the sea to continue their migrations. Whales, however, are so completely adapted to life in the water that they give birth in the ocean.

Social interactions and population dynamics cause some fish, including the pairs of males and females above, to change their appearance—and even their sex. Frogs and toads also reproduce in unique ways; as shown below, their young may be carried in the mouth or on the back.

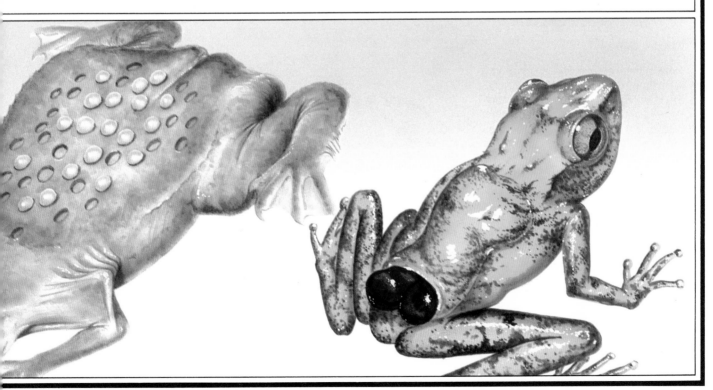

How Do Guppies Reproduce?

In contrast to the drab, light brown coloring of their female counterparts, male guppies sport bold, bright colors. These color patterns serve a vital purpose: During the breeding season, the females allow only the most colorful males to fertilize their eggs.

Such popularity has its pitfalls, however, for the males with the brightest colors attract predators as well as mates. Male guppies that survive in spite of their high visibility are therefore likely to be the strongest and fastest of the species. By mating with such a male, a female increases the chances of passing those desirable traits to the next generation.

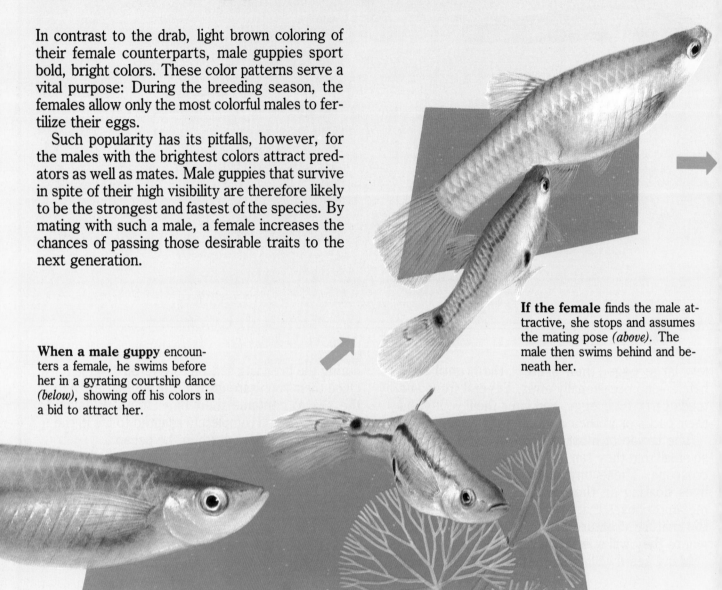

When a male guppy encounters a female, he swims before her in a gyrating courtship dance *(below)*, showing off his colors in a bid to attract her.

If the female finds the male attractive, she stops and assumes the mating pose *(above)*. The male then swims behind and beneath her.

Flashy fish

Among fish species whose colors vary according to sex, the male is usually brighter *(right)*. The male, producing large quantities of sperm, often mates with many females, but a female may mate with few males, or only one. In addition, a brightly colored male tends to attract—and mate with—more females. For these reasons, brightly colored males sire more offspring than less colorful males. This process of natural selection results in body colors that are more and more eye-catching with each succeeding generation.

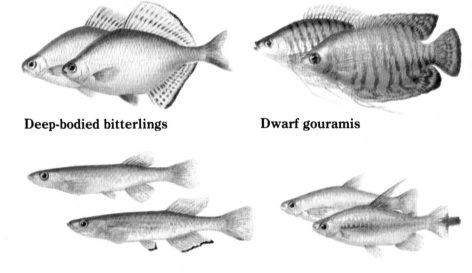

Deep-bodied bitterlings

Dwarf gouramis

African killifish

Congo tetras

Turning upside down *(below)*, the male injects sperm into the female's urogenital opening.

A few weeks after mating, the female bears several dozen tiny guppies. Within six months they become breeding adults *(right)*.

Guppy love

Popular as aquarium fish, guppies have been bred to yield a wide variety of colors. A breeder chooses a male for his colors and pattern, then mates him with different females. The process is repeated with the most colorful males of several generations, producing the rainbow of colors shown here.

Leopard

Tuxedo

King cobra

Albino long fin

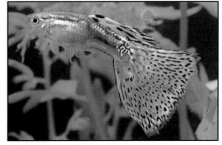

Grass

Why Do Angelfish Change Sex?

Blackspot angelfish spawn in the evening. The male and the female rise from the bottom, releasing their sperm and eggs into the water almost simultaneously *(above)*. The male then moves on to engage another female in spawning.

When the harem's only male dies *(above)*, the largest female begins to assume his appearance. As a first step in this transformation, the female grows larger, but her color—yellow, with a black-edged tail—remains the same.

From female to male

The females of more than 200 fish species can become males when circumstances require. Most of these species are polygynous—that is, a single male usually mates with numerous females. In most cases, the male is larger. As shown at right, the colors and patterns of the sexes often vary.

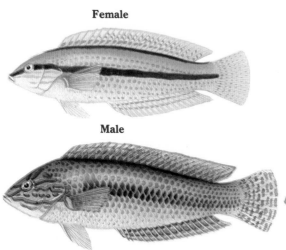

The puddingwife, found along the Atlantic coast from the Florida Keys to Brazil, goes by two names: red wrasse for females, blue wrasse for males.

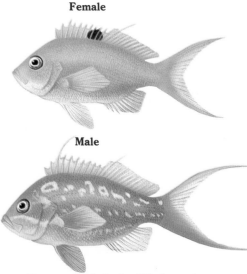

Because of their differing colors, the male and female cherry bass were once thought to be two separate species.

Blackspot angelfish live in groups, or harems, made up of one male and about five females. The male is the largest and strongest member of the group; there is no such thing as a small male blackspot angelfish. When the male dies, the largest female undergoes a sex change, becoming a male. The eggs in its body are absorbed, and the new male begins to produce sperm.

This arrangement is a prime example of hermaphroditism—a reproductive mode in which a fish is capable of functioning as either a male or a female. Because the survival of the harem is jeopardized whenever the single male dies, the blackspot angelfish has evolved a mechanism, illustrated below, that allows the largest female to take on the male's role.

About a week after the male dies, the largest female starts to take on the colors and patterns of a male. The convert begins to show the same courtship behavior toward the females that was displayed by the original male.

Just two weeks after the death of the original male, black stripes on the largest fish indicate that it has completed the sex change. The new male fish now courts the females and fertilizes their eggs with his sperm.

From male to female

In about 40 species, male fish can change their sex to become females. These fish usually live in groups, with the female as the largest member. The next largest is a male, while the other members of the group are immature fish. When the female dies, the male changes its sex and becomes a female; the next-largest fish in the group then matures into a female.

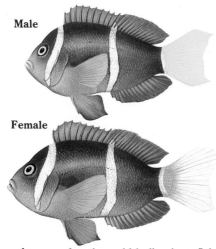

As a male, the gold-belly clownfish has a yellow tail fin *(top)*. After the fish changes its sex to female, the tail fin turns white *(bottom)*.

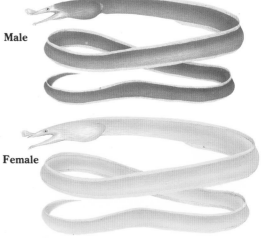

The male moray eel shown at top has a brilliant blue body color. Upon changing into a female, the eel becomes yellow.

67

What Are Hermaphrodites?

Like the blackspot angelfish, the hamlet is an example of a hermaphrodite—an organism that has both male and female reproductive organs (testes and ovaries). Despite this double endowment, the hamlet cannot reproduce on its own. As evening comes on, two hamlets pair off and mate for about 30 minutes; exchanging the male and female sexual roles, they spawn about 10 times before the sun goes down.

● **A two-in-one fish**

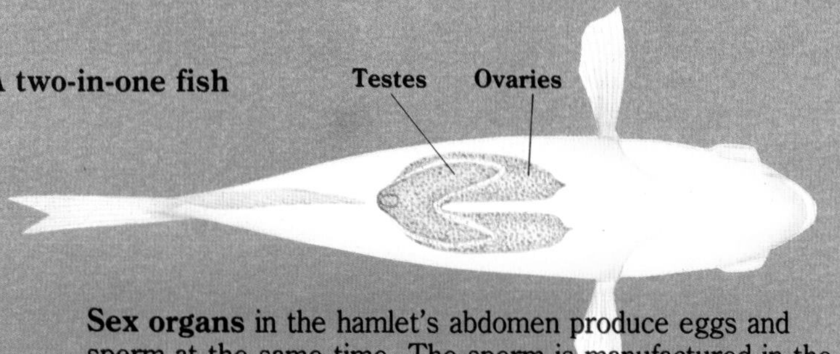

Testes   Ovaries

Sex organs in the hamlet's abdomen produce eggs and sperm at the same time. The sperm is manufactured in the thin, cordlike testes, located beneath the large ovaries.

Ⓐ

Ⓑ

Two hamlets pair off for breeding as sunset nears. One partner takes the role of the female (B), swimming around its mate to persuade it to rise from the ocean floor.

Ⓐ

Ⓑ

The partner in the male role (A) lies over its mate, curving its body to hold the other. The female jerks its body to lay eggs; at the same time, the male releases sperm to fertilize the eggs. The fertilized eggs then drift away.

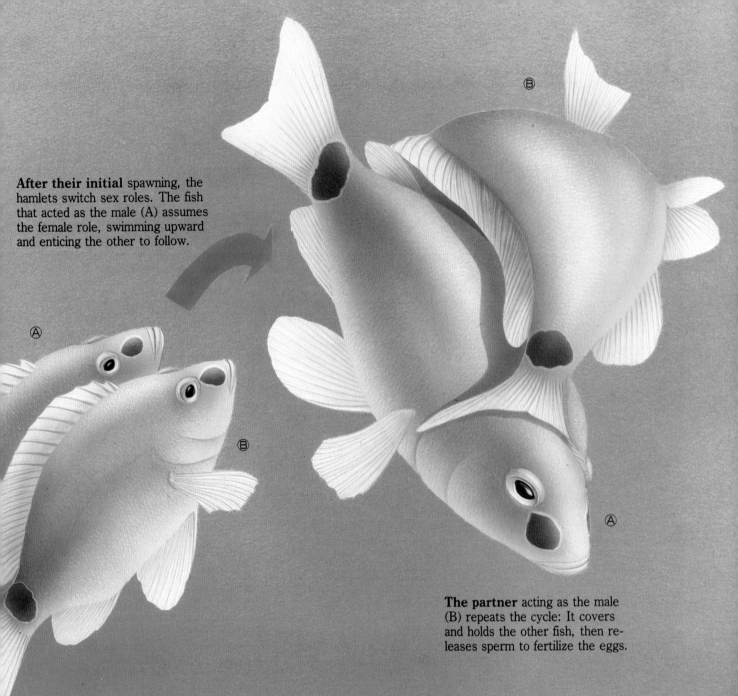

After their initial spawning, the hamlets switch sex roles. The fish that acted as the male (A) assumes the female role, swimming upward and enticing the other to follow.

The partner acting as the male (B) repeats the cycle: It covers and holds the other fish, then releases sperm to fertilize the eggs.

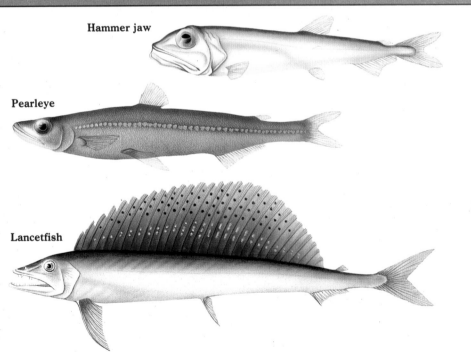

Deep-sea hermaphrodites

Many hermaphroditic fish are pelagic, meaning they inhabit the open seas. In these vast expanses, where potential partners are few and far between, the fishes shown at right have evolved to make the most of their meetings with members of the same species. Each is equipped with both eggs and sperm, so mating can occur at every encounter.

Hammer jaw

Pearleye

Lancetfish

What Are Sexual Parasites?

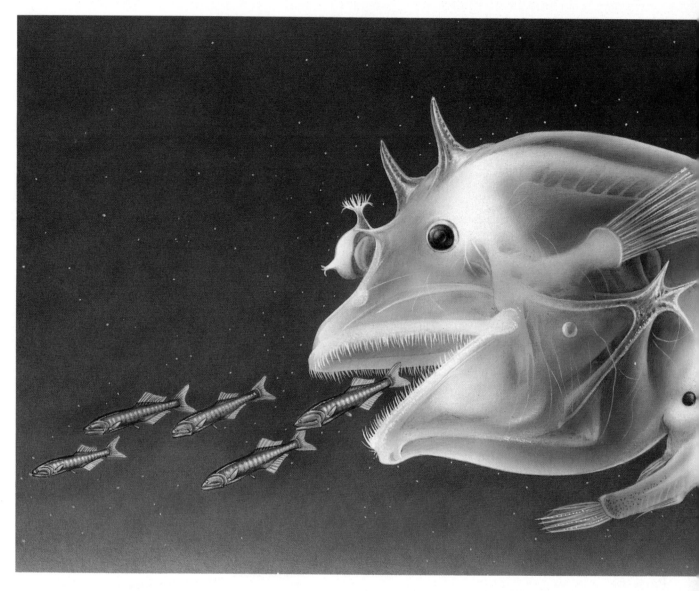

**Once you have found her,
never let her go**

The deep-sea anglerfish is the biggest of its kind. Females grow to be more than a yard long and can produce millions of eggs in a single spawning. The males, which live alone until maturity, reach a length of only 4 to 6 inches.

As shown at right, the early development of both sexes is roughly the same. At a length of about ⅙ inch *(near right),* each has a balloonlike body caused by a gelatin layer under the skin. The female then develops a luminous lure; she uses this "fishing rod" to angle for, or attract, prey. The male grows pincerlike teeth, which will grip the skin of the female. When their growth is complete, the female is 12 times longer than the male.

Female

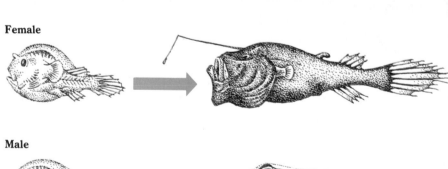

Male

Before metamorphosis, each fish is ⅙ inch long. The lure has begun to form above the female's eyes.

As the fish mature, the female's luminous organ becomes fully developed, and the male acquires teeth.

In the perpetual dark that prevails 2 miles below the surface of the sea, finding a mate requires blind luck. Because of this difficulty, the male anglerfish literally gloms onto the first female he finds. From that day forward, the tiny male lives as a parasite on the body of the much larger female, taking his food from her bloodstream. In time their bodies fuse together, forming a sort of two-body hermaphrodite. Although this arrangement primarily benefits the male, it also frees both sexes from constantly seeking out new breeding partners whenever it is time to mate.

Resembling an ocean ghost, a large female ceratoid anglerfish carries two smaller males as she chases a meal 10,000 feet beneath the surface. The female lives in a host-parasite relationship with up to three males.

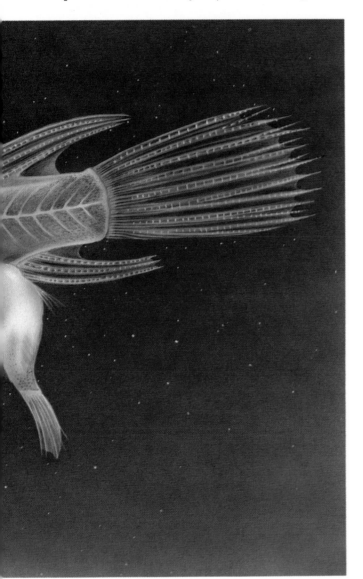

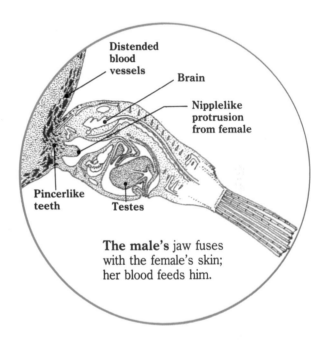

Distended blood vessels

Brain

Nipplelike protrusion from female

Pincerlike teeth

Testes

The male's jaw fuses with the female's skin; her blood feeds him.

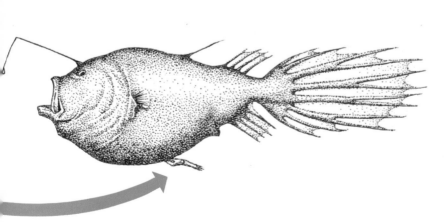

An adult female swims with a male in tow, his teeth attached to her. Eventually the male's jaw will coalesce with the female's skin, connecting them permanently.

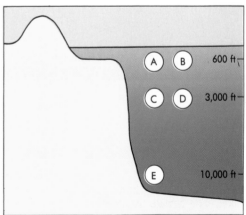

		600 ft
A	B	
		3,000 ft
C	D	
E		10,000 ft

Released in the deep, anglerfish eggs float to the surface, where the young fish, or fry, feed on plankton (A and B). Metamorphosis occurs at about 3,000 feet (C and D). The adults dwell in the depths (E).

Why Do Fish Lay So Many Eggs?

Most marine fish spawn a stupendous number of eggs, but the giant sunfish *(right)* is the champion of them all: A single female releases as many as 30 million eggs at a time. If every one of these eggs grew to maturity, the oceans would soon be wall-to-wall sunfish. Because the eggs are left unprotected after spawning, however, most of them succumb to disease, starvation, or attacks by predators. Colossal quantities of eggs are therefore required to guarantee that some will live to become parents themselves. Fish that take care of their eggs, by contrast, can lay them in smaller numbers, because a greater proportion will survive to adulthood.

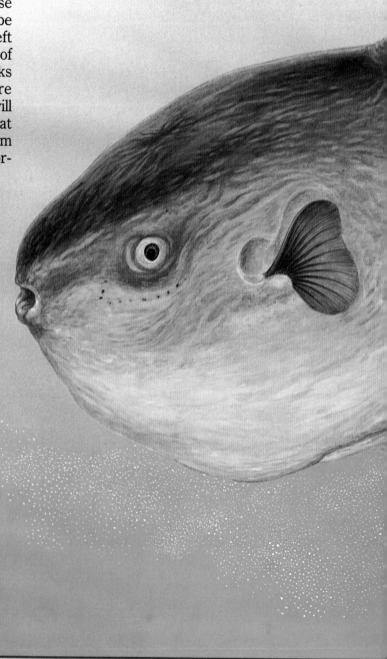

The sea-dwelling scorpionfish *(left)* bears about 10,000 live young, called fry, each ⅛ inch long. The fry drift on the surface, feeding on plankton, until they are ¾ inch long. They then settle to the bottom to live as adults.

The livebearing mosquitofish bears up to 100 fry after spawning. Each of the young fish is less than ⅓ inch long.

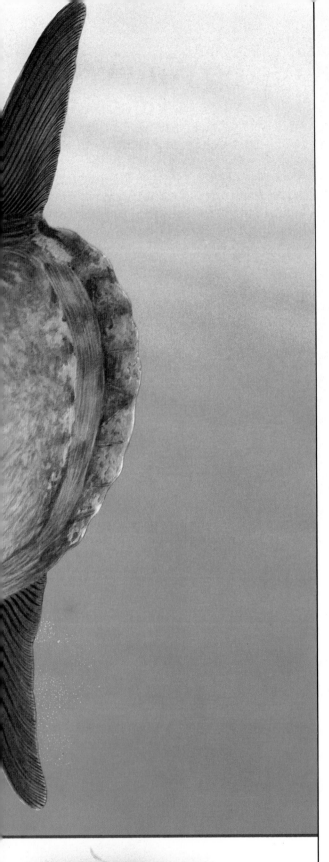

The spawning spectrum

A Japanese eel lays more than a million eggs, each the size of a pinhead, which hatch in the ocean. The tiny fry travel to rivers and swim upstream.

The female bluefin tuna, up to 6 feet long, carries as many as 5 million tiny eggs. The newly hatched fry are less than ⅛ inch long.

Japanese pilchards *(left)* spawn tens of thousands of eggs; only 1 in 1,000 survives two months.

The sticky eggs of the Japanese dace cling to river stones. The eggs hatch a week later.

The freshwater floating goby lays about 1,000 yellow eggs in shells or under rocks. The male guards them; they hatch within 10 days.

A female bitterling uses her tubular ovipositor to lay about 10 eggs in the shell of a living bivalve such as the freshwater mussel at right.

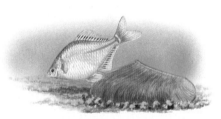

The Egyptian mouthbrooder *(left)* spawns about ten ¼-inch-long eggs. The female holds the eggs in her mouth to protect them during incubation.

The cartilaginous ratfish lays its eggs in spindle-shaped cases up to ¾ inch long. Each case holds two eggs.

Can Fish Lay Eggs out of the Water?

Despite the best efforts of their parents, eggs and young fish are hard to protect from predators. One way to keep eggs safe is to lay them where other fish cannot go. The splashing tetra, native to South America, has found just such a place: the undersides of leaves that overhang the surface of the water by an inch or two.

After locating a leaf suitable for their eggs, the male and female splashing tetras both leap into the air. Clinging briefly to the leaf, they deposit a sticky mass of fertilized eggs. The pair then drop back into the water. The male remains in the area beneath the eggs, splashing water with his tail to keep them moist until they hatch.

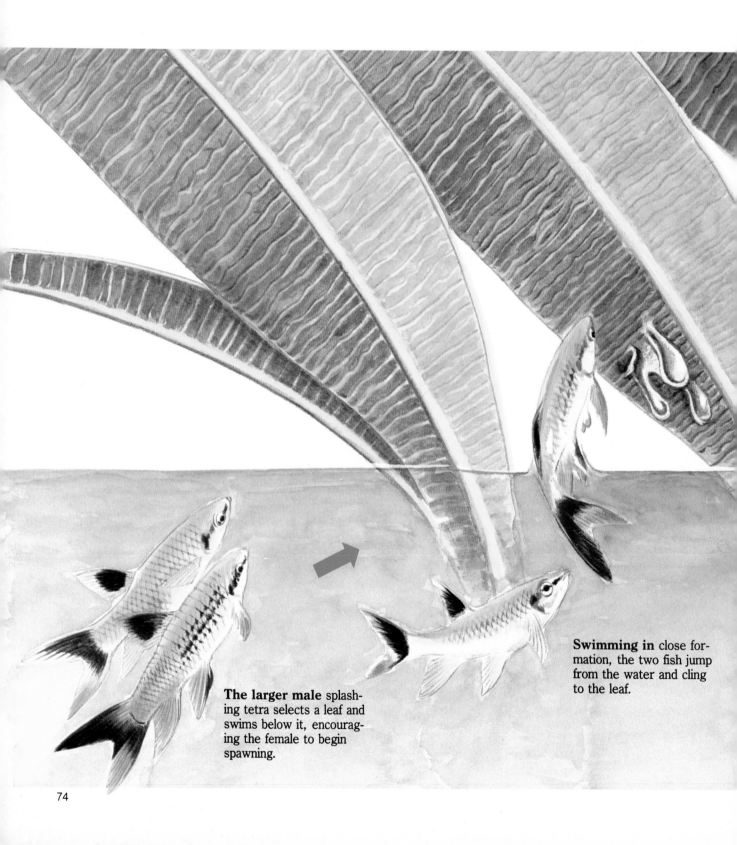

The larger male splashing tetra selects a leaf and swims below it, encouraging the female to begin spawning.

Swimming in close formation, the two fish jump from the water and cling to the leaf.

Although the splashing tetra is the only fish that leaves the water to spawn, some related species are equally dedicated to protecting their eggs. One species, the pyrrhulina *(right)*, lays its eggs on the barely submerged leaves of water plants, where the eggs are less likely to be discovered by hungry predators. Another kind of fish, the red-spotted copeina *(far right)*, attaches its eggs to the riverbed, where the male watches over them until they hatch.

To deposit fertilized eggs on leaves just below the surface, a mating pair of pyrrhulinas glide over the leaf together, releasing their eggs and sperm at the same time.

Swimming along the bottom side by side to coordinate the discharge of their eggs and sperm, two red-spotted copeinas spawn on a streambed. The male stays in the area to protect the fertilized eggs.

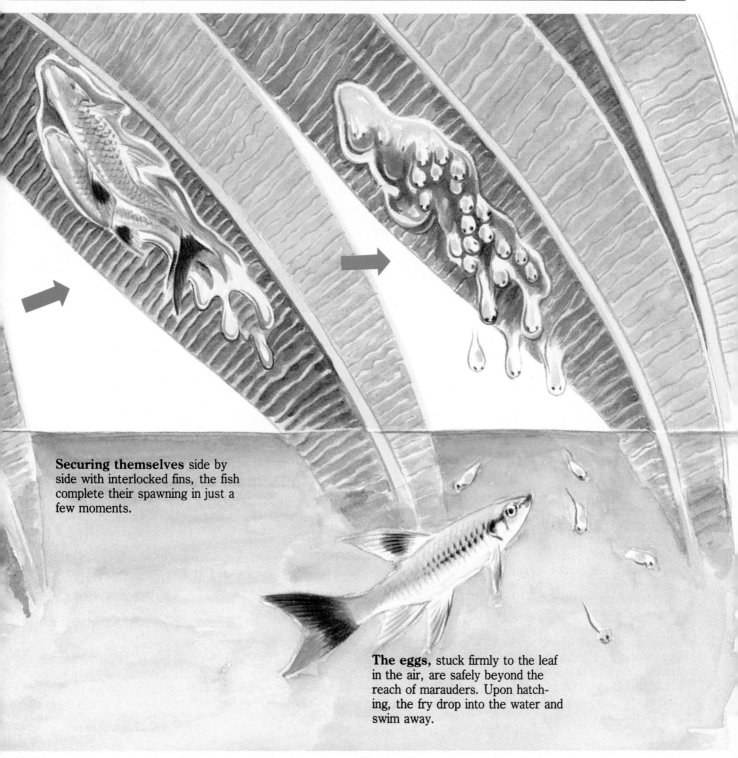

Securing themselves side by side with interlocked fins, the fish complete their spawning in just a few moments.

The eggs, stuck firmly to the leaf in the air, are safely beyond the reach of marauders. Upon hatching, the fry drop into the water and swim away.

How Do Sharks Reproduce?

Some shark species are oviparous—they produce eggs that hatch after being expelled from the body. But most sharks are ovoviviparous (their eggs hatch within the body) or viviparous (the young are born alive).

The reproductive systems of these live-bearing sharks are highly advanced. Dogfish sharks, for example, produce a placenta that nourishes the fetus during the last part of the pregnancy. In some species, the fetus takes its nutrition from the mother for almost the entire gestation period, which lasts eight to nine months. Each fetus develops completely within the mother and is born fully formed. A newborn shark may measure up to 3 feet long.

Fertilized eggs and live births

Egg-laying species are in the minority among sharks. The eggs, fertilized before laying, are released in an egg case and take up to a year to grow to maturity inside it. In some species of cat sharks and nurse sharks, among others, the embryo begins to develop within the mother's body; when it reaches a certain size, the embryo—safely enclosed in a pillow-shaped case—is released into the water. Tendrils curling from the corners of the case get tangled in seaweed or coral, anchoring the case while the embryo develops.

A young swell shark leaves the egg case where it had been growing for months. Inside the case, the shark was nourished by a yolk sac beneath its abdomen.

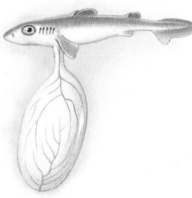

Though nourished by a yolk sac (above), a spiny dogfish fetus develops fully within the mother. Its two-year gestation is among the longest of any vertebrate.

A requiem shark,
one of about 200 live-
bearing species, gives
birth to a fully formed
youngster.

The apparatus of propagation

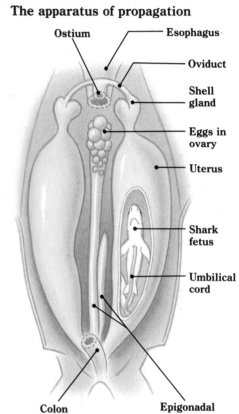

Ostium — Esophagus
Oviduct
Shell gland
Eggs in ovary
Uterus
Shark fetus
Umbilical cord
Colon — Epigonadal organ

Mature eggs from the ovaries enter the
oviduct; in the shell gland, the eggs are
fertilized by sperm from a male shark. In
oviparous sharks, an egg case forms; in
livebearing sharks, a temporary membrane
envelops the egg. The egg then passes to
the uterus. In some species, walls known
as septa keep multiple fetuses separate.

When sibling rivalry goes too far

In some livebearing sharks, the
eggs hatch while they are still
inside the mother. The newly
hatched embryos then sustain
themselves by eating subsequent
eggs as they descend from the
ovaries. Even an embryo that
manages to hatch from its egg
has no guarantee of survival, for
larger embryos frequently canni-
balize their younger siblings.
Though brutal, this arrangement
is effective: The surviving fetuses
may grow to be 3 feet or longer
before birth.

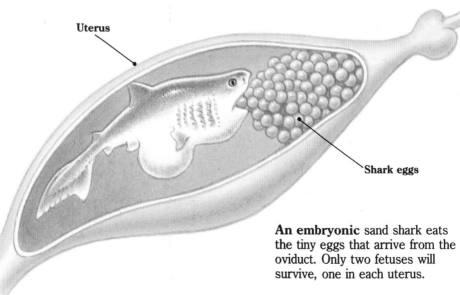

Uterus

Shark eggs

An embryonic sand shark eats
the tiny eggs that arrive from the
oviduct. Only two fetuses will
survive, one in each uterus.

How Do Fish Raise Their Young?

Most fish simply spawn their eggs and then leave them to their fate. Some species, however—including the ones pictured here—take an active role in caring for their eggs and hatchlings.

Each species has a different means of protecting its young. Some fish carry their eggs in their mouths, while others shelter their eggs in nests like those of birds. Certain species supervise their young as they swim about in shoals, or schools, defending them from larger fish and other threats. Ordinarily, only the male or the female watches over the young; on occasion, though, both parents—and even some of the offspring—may share in the task.

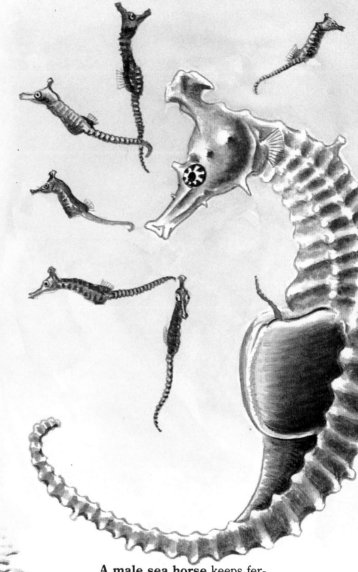

A male sea horse keeps fertilized eggs in a pouch on his abdomen until they hatch.

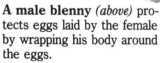

A male blenny *(above)* protects eggs laid by the female by wrapping his body around the eggs.

The female floating goby lays eggs that stick to the top of a rock crevice *(right);* the male then stands guard underneath the eggs until they hatch.

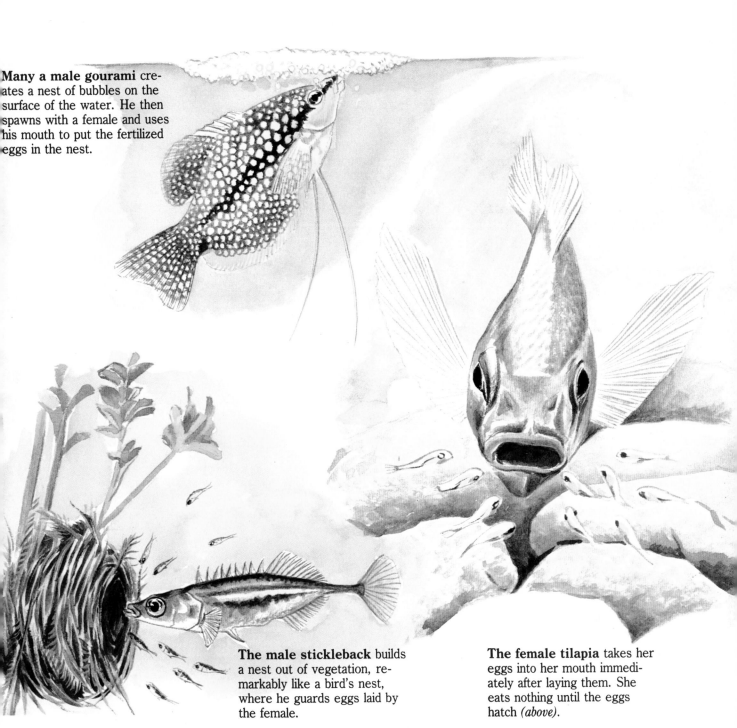

Many a male gourami creates a nest of bubbles on the surface of the water. He then spawns with a female and uses his mouth to put the fertilized eggs in the nest.

The male stickleback builds a nest out of vegetation, remarkably like a bird's nest, where he guards eggs laid by the female.

The female tilapia takes her eggs into her mouth immediately after laying them. She eats nothing until the eggs hatch *(above)*.

All in the family

Hundreds of species of cichlids live in the great rift lakes of Africa. In many of these species, both the male and the female watch over their eggs and newly hatched offspring. In time, young cichlids that hatched earlier but are still schooling with their parents begin to take on some of the caretaking duties. These older offspring may continue in this protective role until they reach breeding age themselves.

What Are Mouthbrooders?

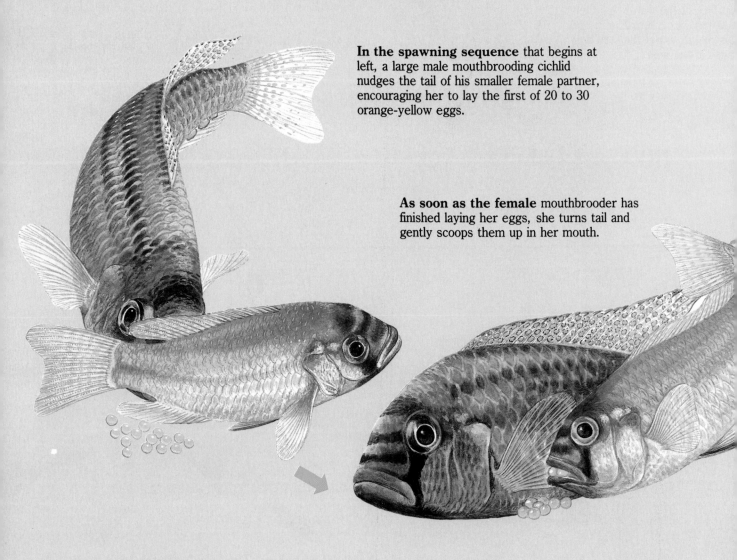

In the spawning sequence that begins at left, a large male mouthbrooding cichlid nudges the tail of his smaller female partner, encouraging her to lay the first of 20 to 30 orange-yellow eggs.

As soon as the female mouthbrooder has finished laying her eggs, she turns tail and gently scoops them up in her mouth.

Father broods best

In some species of mouthbrooders, such as the reef-dwelling blackspot cardinalfish shown below, the male incubates the eggs. As soon as the female releases eggs, the male takes them in his mouth.

A female cardinalfish spawns an egg sac that contains thousands of eggs. To prevent predators from reaching the eggs first, the male waits just behind and below the female.

The male mouthbrooder captures the egg sac in his protruding jaw, where he coddles the eggs during the week or so it takes them to hatch. Once the fry emerge from the eggs, the male abandons them to grow up on their own.

Of all the strategies that fish have evolved to protect their young from predators, the oddest may be mouthbrooding—the incubation of eggs inside the mouth of a parent fish. Both freshwater and saltwater species engage in this behavior, and either males or females may carry the eggs.

Although the parent fish cannot feed while the eggs develop, the embryonic fish feast on protein contained in a yolk sac. After one to two weeks, having grown large enough, the hatchlings leave the haven of the parent's mouth to brave the hazards of river, lake, or sea.

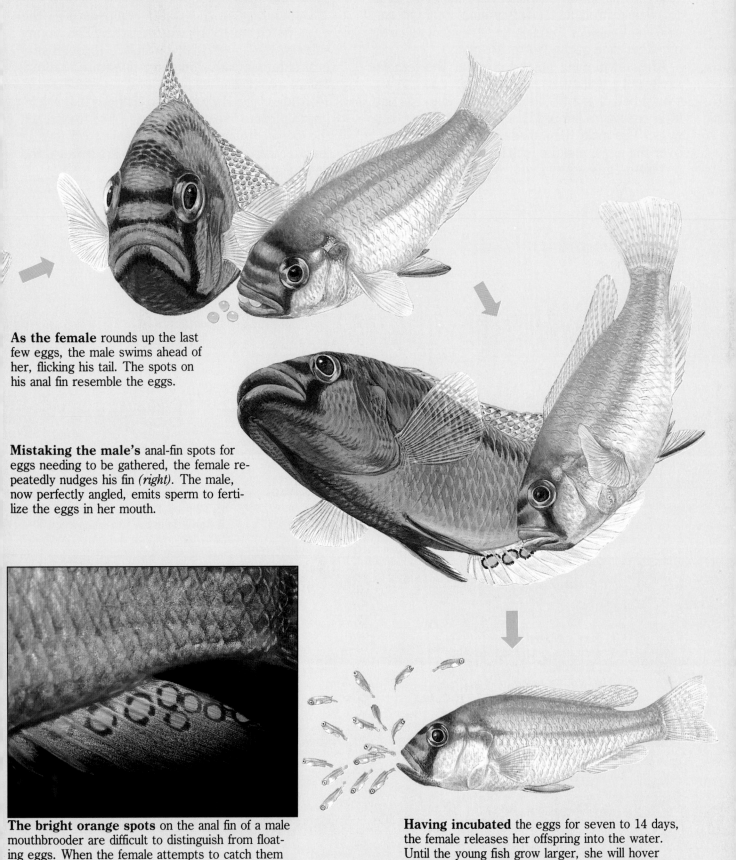

As the female rounds up the last few eggs, the male swims ahead of her, flicking his tail. The spots on his anal fin resemble the eggs.

Mistaking the male's anal-fin spots for eggs needing to be gathered, the female repeatedly nudges his fin *(right)*. The male, now perfectly angled, emits sperm to fertilize the eggs in her mouth.

The bright orange spots on the anal fin of a male mouthbrooder are difficult to distinguish from floating eggs. When the female attempts to catch them in her mouth, fertilization takes place.

Having incubated the eggs for seven to 14 days, the female releases her offspring into the water. Until the young fish grow larger, she will hover nearby, letting them retreat into her mouth whenever danger threatens.

81

How Do Salmon Develop?

Salmon go to great lengths—literally—in order to reproduce. When the Atlantic salmon gets ready to spawn, for example, it swims more than 1,000 miles from its feeding grounds near Greenland and Labrador to reach the stream of its birth on either side of the North Atlantic.

After swimming up the stream, the female seeks out a shallow area where clean water flows over a bed of gravel. Using her tail, she digs out little basins, called redds, in which she lays her eggs. The male then sheds his milt, or sperm, over the eggs, and the female quickly covers the fertilized eggs with gravel. After spawning, the weakened adults drift downstream; some die before reaching the ocean, but many survive to migrate and spawn again.

Depending on the water temperature, the eggs hatch up to 16 weeks later. The young salmon linger near their birthplace through several stages of development, illustrated below, before riding the stream's swift current to the sea. Internally, they undergo changes that will allow the fish to live in salt water. Once they reach the ocean, the salmon may grow to 3 feet in length before heading back upstream to begin the spawning cycle anew.

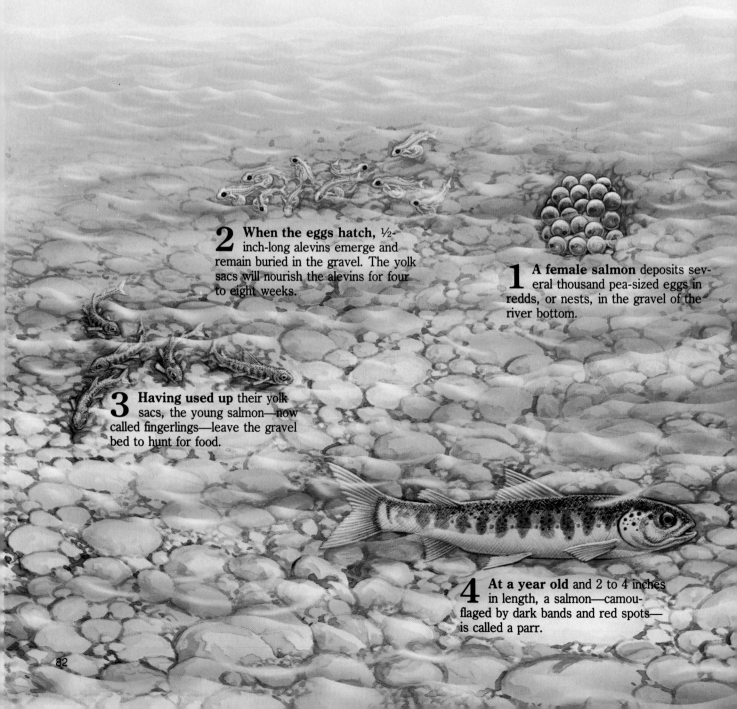

2 **When the eggs hatch,** ½-inch-long alevins emerge and remain buried in the gravel. The yolk sacs will nourish the alevins for four to eight weeks.

1 **A female salmon** deposits several thousand pea-sized eggs in redds, or nests, in the gravel of the river bottom.

3 **Having used up** their yolk sacs, the young salmon—now called fingerlings—leave the gravel bed to hunt for food.

4 **At a year old** and 2 to 4 inches in length, a salmon—camouflaged by dark bands and red spots—is called a parr.

During the years that it lives in the sea, a mature salmon is strong and silvery, with abundant body fat and muscle. As it swims upstream to spawn, leaping as high as 10 feet to clear waterfalls, the salmon uses up its reserves of fat. The silver gives way to a dark brown or bronze background, mottled with red and orange.

6 **Despite the rigors** of their upstream spawning run, many Atlantic salmon survive to return two or more times.

5 **After another year or so** of maturation, the young salmon loses its parr marks, which hid it well in the shallow river, and becomes a silvery smolt—a color scheme better suited to its next destination, the open ocean. The smolt, 4 to 8 inches long, embarks on a journey to the sea; along the way, its body adapts so the fish can survive in salt water.

Do Some Fish Trick Others into Raising Their Young?

Among birds, cuckoos and cowbirds are infamous for sneakily laying their eggs in the nests of other species. Certain fish, too, avoid incubating their own eggs by tricking others into doing the work for them. The cuckoo catfish, a native of Africa's Lake Tanganyika, plays this ruse on mouthbrooding cichlids.

While the cichlids are in the midst of spawning, a pair of catfish swoop down upon them and swallow some of their eggs. Then, swimming past the female cichlid, the female cuckoo catfish releases her own eggs and the male issues a cloud of his sperm. Confused by these actions, the unwitting female cichlid mistakenly takes the catfish eggs into her mouth and incubates them along with her own eggs.

2 **Maturing** rapidly, the catfish eggs hatch before those of the cichlid. At first, the catfish fry live on their own yolk sacs.

1 **A female** cichlid holds a clutch of eggs in her mouth. Unknown to her, the clutch may include eggs of a cuckoo catfish.

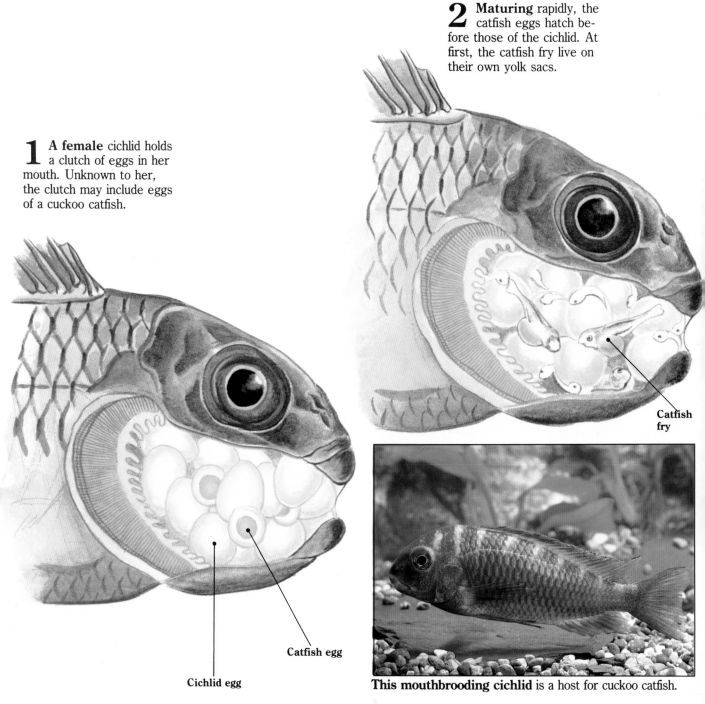

Catfish fry

Catfish egg

Cichlid egg

This mouthbrooding cichlid is a host for cuckoo catfish.

A cuckoo catfish probes the bottom with its barbels.

4 **As they grow,** the catfish fry prey on the smaller cichlids. This parasitic relationship gives the catfish fry protection and a rich supply of food.

3 **By the time** the cichlid eggs begin to hatch, the catfish fry have consumed their own yolk sacs. They may then start to devour the cichlid fry.

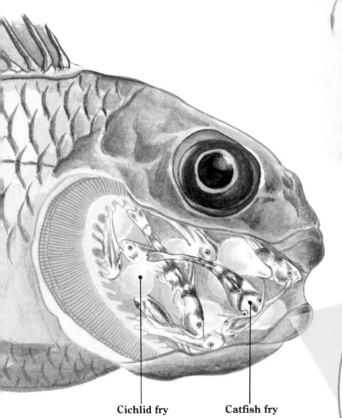

Cichlid fry   Catfish fry

Catfish fry

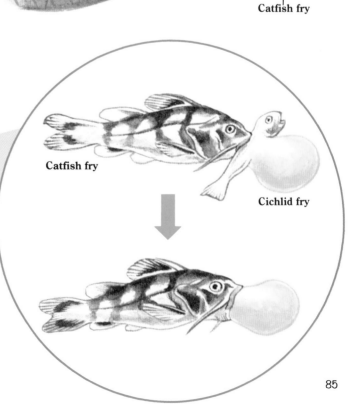

Catfish fry

Cichlid fry

Cichlid eggs are covered with a tough skin that keeps them from being eaten by catfish fry. Upon hatching, however, the soft larval cichlids and their nutrient-filled yolks make easy targets for the hungry interlopers *(right)*.

Which Fish Feed Their Young?

Whereas mammals nurse their young and virtually all birds bring food to their chicks, the vast majority of fish are indifferent parents. Most newly hatched fish must fend for themselves from the moment they leave their eggs.

There are a number of interesting exceptions to this rule, however. The brilliantly colored South American discus, shown below and at right, sustains its young with a protein-packed mucus that seeps from its skin. In Africa's Lake Nyasa, male and female catfish dig a muddy nest where they provide food for their offspring. And the females of some livebearing species, which produce living young rather than eggs, feed their larvae a nourishing substance secreted from their reproductive tracts.

A male discus fertilizes a group of 100 to 200 eggs that the female has just deposited on a piece of driftwood.

The male and female take turns fanning the eggs with their fins. The water flow provides oxygen and keeps the eggs free of disease.

A thoroughly modern catfish couple

Some bottom-dwelling catfish of Lake Nyasa diligently tend to the needs of their young. Both adults guard the eggs until the fry emerge, at which point the parents take turns feeding their offspring. The male swims off in search of vegetation, small fish, or other creatures. After cramming these tiny morsels into his mouth, he returns to the nest and expels the food through his gills. The female then lays a batch of unfertilized eggs for the young to ingest. Few other fish species lavish such care on their hatchlings.

Nuzzling their male parent, hungry catfish fry retrieve bits of food ejected from his gills.

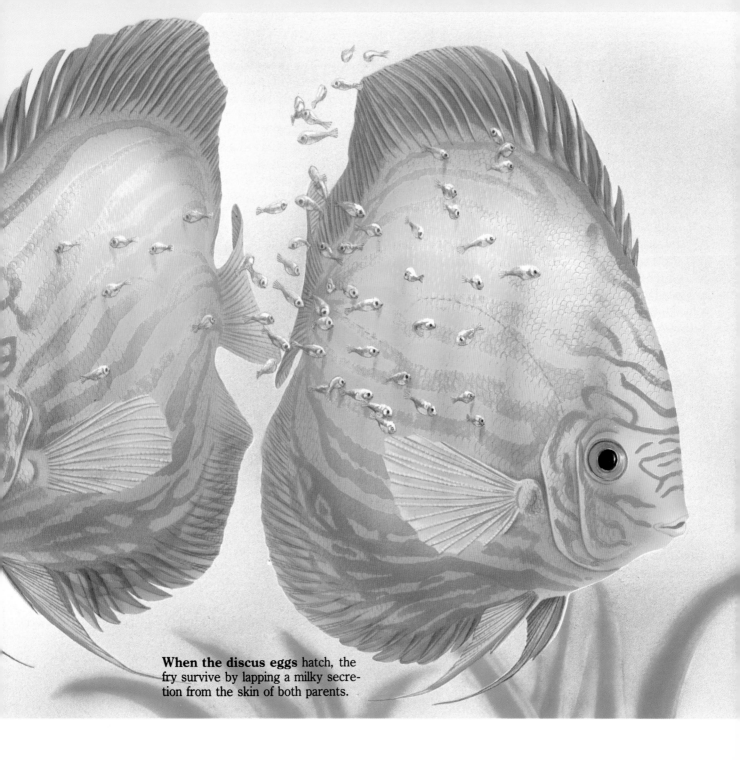

When the discus eggs hatch, the fry survive by lapping a milky secretion from the skin of both parents.

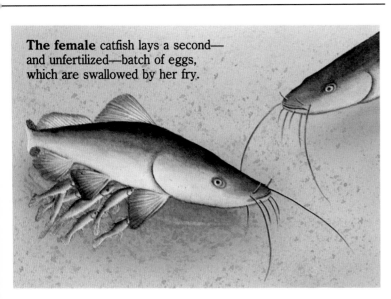

The female catfish lays a second—and unfertilized—batch of eggs, which are swallowed by her fry.

A female surffish, native to the northwest Pacific Ocean, swims with her fry. The eggs are fertilized and hatched in the womb; the fry remain there for a time, nourished by a liquid made in the ovaries.

What Is Metamorphosis?

Metamorphosis is a growth process so dramatic that the larvae and adults of the same creature are easily mistaken for different species. The process may enable a young animal to take on a protective form that can be "shed" when no longer needed, or it may allow an animal to survive in two separate environments during its life.

The ocean sunfish shown at right, for example, is highly vulnerable after hatching. But it soon develops a coat of sharp spines, five of which grow into long spikes designed to repel predators. By the time it is fully grown, the sunfish has no need of body armor: The 3-inch-thick layer of gristle beneath the skin of an adult has been known to stop rifle bullets.

Metamorphosis helps amphibians such as frogs *(below)* evolve from aquatic gill breathers to land dwellers who breathe with lungs. The young frogs, called tadpoles, resemble fish in that they use a tail and fins for swimming; the adult frogs are closer to true land vertebrates in that they use two pairs of limbs for jumping.

From mite to monster

 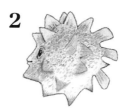

Living in shallow areas of tropical oceans, the ocean sunfish *(right)* is exposed to many threats. Shortly after it hatches (1), the ⅛-inch-long sunfish develops a covering of spines (2). By the time it reaches ½ inch, its armament has been simplified to five spikes (3). The spikes dwindle as the sunfish grows to an inch long (4). Eventually, they are replaced by a smooth covering of scales (5). An adult sunfish may ultimately reach a length of 11 feet and weigh as much as a ton.

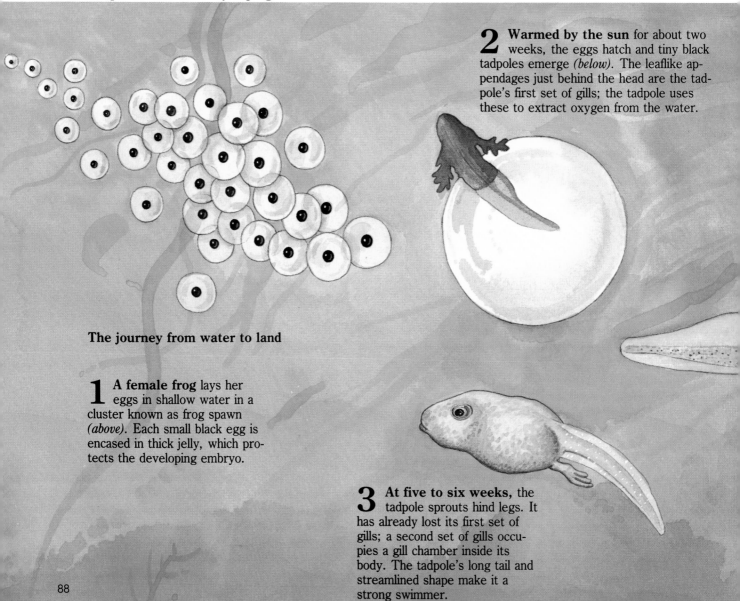

2 **Warmed by the sun** for about two weeks, the eggs hatch and tiny black tadpoles emerge *(below)*. The leaflike appendages just behind the head are the tadpole's first set of gills; the tadpole uses these to extract oxygen from the water.

The journey from water to land

1 **A female frog** lays her eggs in shallow water in a cluster known as frog spawn *(above)*. Each small black egg is encased in thick jelly, which protects the developing embryo.

3 **At five to six weeks,** the tadpole sprouts hind legs. It has already lost its first set of gills; a second set of gills occupies a gill chamber inside its body. The tadpole's long tail and streamlined shape make it a strong swimmer.

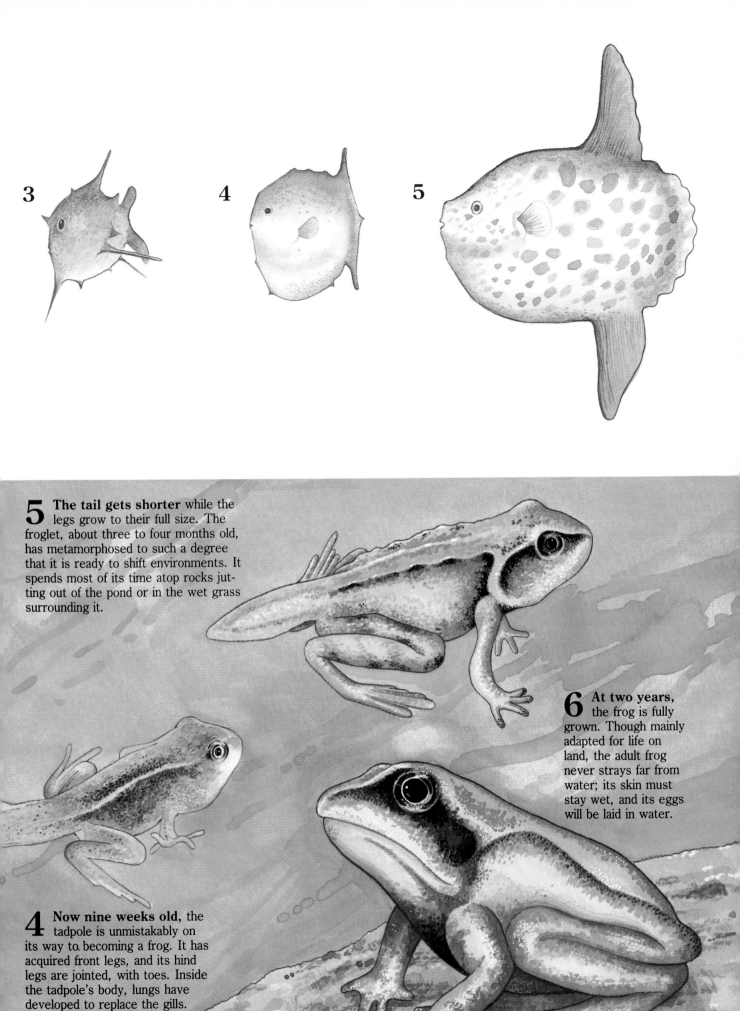

3

4

5

5 **The tail gets shorter** while the legs grow to their full size. The froglet, about three to four months old, has metamorphosed to such a degree that it is ready to shift environments. It spends most of its time atop rocks jutting out of the pond or in the wet grass surrounding it.

6 **At two years,** the frog is fully grown. Though mainly adapted for life on land, the adult frog never strays far from water; its skin must stay wet, and its eggs will be laid in water.

4 **Now nine weeks old,** the tadpole is unmistakably on its way to becoming a frog. It has acquired front legs, and its hind legs are jointed, with toes. Inside the tadpole's body, lungs have developed to replace the gills. The creature must therefore surface often to gulp air.

How Do Frogs and Toads Reproduce?

Most frogs and toads shed their eggs and then abandon them to the vagaries of nature. Because the eggs make a tasty meal for water beetles, fish, and other predators, a clutch of 20,000 eggs may yield very few surviving young.

A number of frog and toad species, however, are much more active—and protective—parents. Rather than laying thousands of eggs to ensure that a few survive, they produce just a handful of eggs and guard them carefully after they hatch. Some of the ingenious strategies these species have developed for improving their odds of reproducing are shown here.

A land-based male European midwife toad carries fertilized eggs on his hind legs to keep them moist. At hatching time, he immerses himself in water, and tadpoles emerge.

Crouching in a treetop pool of rainwater caught by a bromeliad plant, a South American pygmy marsupial frog releases the tadpoles that were stuck to her back by mucus.

Four tadpoles cling to their mother, an arrow-poison frog. The eggs are laid on plants or the ground, where they hatch; the tadpoles may then be carried to the treetops to mature.

A froglet crawls from the mouth of its mother, an Australian gastric-brooding frog. The female swallows her own eggs, then incubates them in her stomach; her digestion is suspended during this time. After the eggs hatch and develop, the mother gives birth to the frogs through her mouth.

A male Darwin's toad watches over his offspring. The male carries fertilized eggs in vocal sacs—long pouches located on his sides—until the young toads emerge from his mouth.

Having incubated for an average of 107 days in chambers embedded in their mother's back, tiny Surinam toads break through membranes of skin to swim free.

The female aquatic Carvalho's pipa, a relative of the Surinam toad, carries her fertilized eggs on her back, where they were spread by a web-footed male. The tadpoles hatch in 14 to 28 days and promptly swim away *(far right)*.

How Do Fur Seals Breed?

Able to use all four limbs for walking, fur seals—descendants of the same creatures that gave rise to bears and dogs—feel right at home on dry land. Yet fur seals also undertake one of the longest sea migrations known, traveling some 6,000 miles to breed.

The fur seals spend the winter as far south as San Diego, California. In late spring they return to their traditional rookeries, or breeding colonies, on the Pribilof Islands of Alaska. There, each adult male fur seal gathers a harem of females, who give birth to single pups conceived the previous year.

Bull fur seals, measuring up to 7 feet long and weighing up to 600 pounds, reach the rookery in late April or May. They fight one another for territory and for mating rights with the females, which arrive in mid-June and July.

Breeding habits of true seals

The true seals at right differ from the fur seals above in that their rubbery rear flippers are of no use on land. These same flippers, however, make true seals among the best swimmers and divers of any mammal; they can dive repeatedly to 300 feet. One northern elephant seal dived to the astonishing depth of 2,900 feet—below the range of some whales. True seals spend most of their time playing, lolling, or hunting in the water. They mate and breed as close as possible to the sea.

Pregnant females of the harbor seal variety shown above favor isolated Arctic pack ice. After giving birth in May, the female nurses her calf for a month.

Snug in a den she has clawed in the Arctic ice, a female ringed seal guards her newborn calf against polar bears and other predators.

A bull fur seal weighs up to four times more than each of the two to 50 grayish females that make up his harem.

Not long after reaching the rookery, a female fur seal delivers a single pup. She mates again by early August or so. The new embryo will gestate for one year, and the pup will be born the next summer, often in the same rookery. The fur seals swim south in the autumn.

regarious harbor seals of the variety shown above bask in roups on rocky coasts of North America. They mate in the ater, usually in September. Unlike many other true seals, arbor seals are often born with the same coloration as adults.

Easy to recognize by the males' drooping snouts, elephant seals crowd an island breeding ground off California. Elephant-seal calves, born in the spring, risk being crushed by their fathers, who weigh up to 3½ tons.

Where Do Gray Whales Migrate?

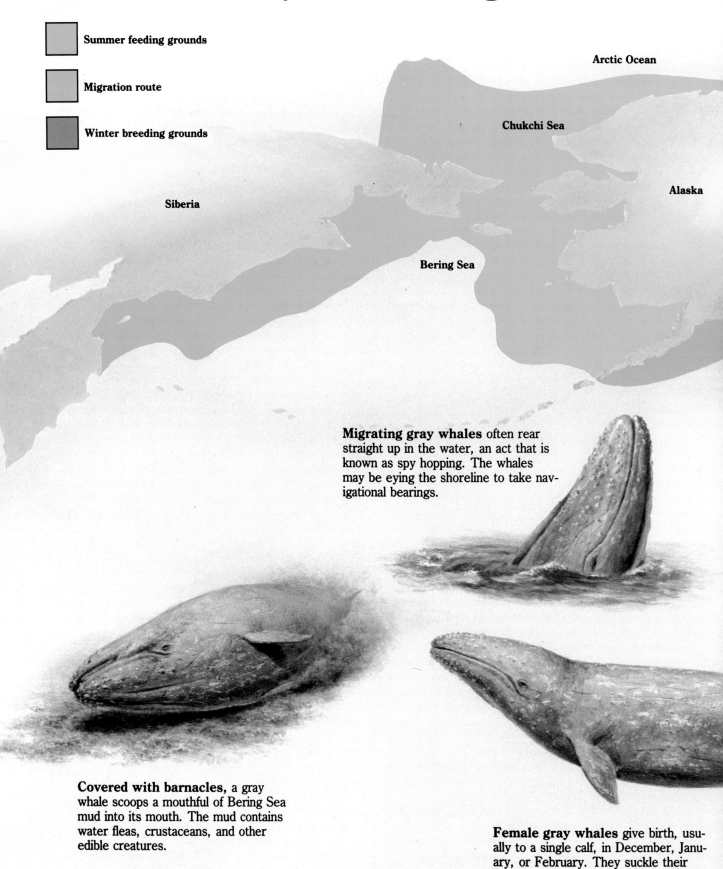

- Summer feeding grounds
- Migration route
- Winter breeding grounds

Arctic Ocean

Chukchi Sea

Siberia

Alaska

Bering Sea

Migrating gray whales often rear straight up in the water, an act that is known as spy hopping. The whales may be eying the shoreline to take navigational bearings.

Covered with barnacles, a gray whale scoops a mouthful of Bering Sea mud into its mouth. The mud contains water fleas, crustaceans, and other edible creatures.

Female gray whales give birth, usually to a single calf, in December, January, or February. They suckle their newborn offspring in lagoons in the Gulf of California.

California gray whales, some of which grow to 50 feet in length and weigh more than 36 tons, have been hunted to the brink of extinction. Yet they still make a yearly pilgrimage of 10,000 miles through the Pacific Ocean.

In summer, the gray whales search for food near the Arctic Circle in the Bering and Chukchi seas. Late in the fall, pregnant females, traveling alone and at high speed, head south for the Gulf of California. They are followed by small groups of males, females with calves, and childless females. The whales cover 60 to 70 miles a day.

Winters are devoted to giving birth and mating in the protected lagoons off Baja California. Come springtime, the newly pregnant females lead off the return trip north, followed by adult males and unbred females. Finally, the mothers and their newborn calves bring up the rear.

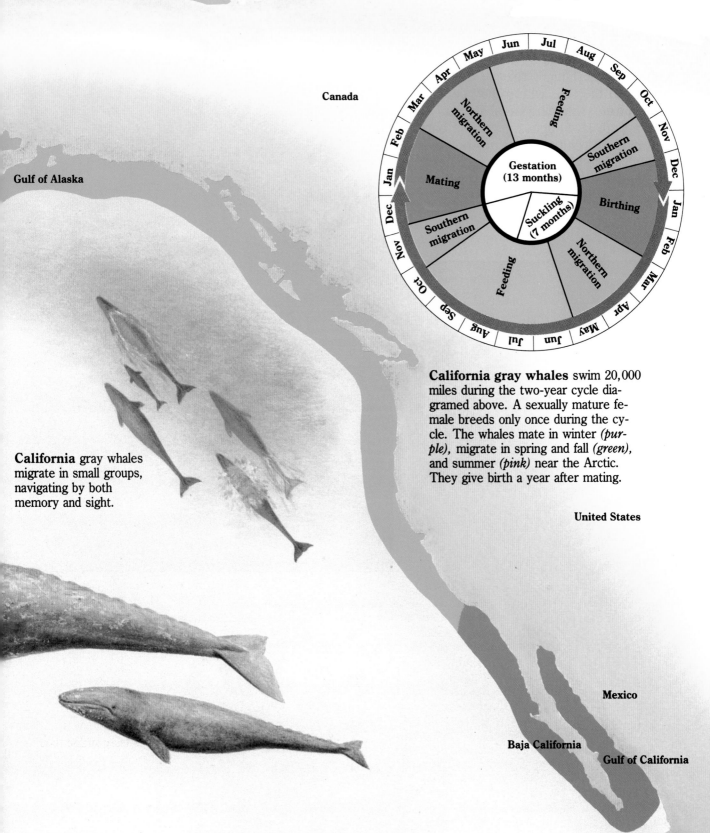

California gray whales migrate in small groups, navigating by both memory and sight.

California gray whales swim 20,000 miles during the two-year cycle diagramed above. A sexually mature female breeds only once during the cycle. The whales mate in winter *(purple)*, migrate in spring and fall *(green)*, and summer *(pink)* near the Arctic. They give birth a year after mating.

United States

Mexico

Baja California

Gulf of California

Gulf of Alaska

Canada

How Do Crabs Develop?

Crabs reproduce sexually. Depending on the species—there are 6,000 worldwide—a female crab may lay as few as 5,000 or as many as 3 million eggs in a single batch. Many female freshwater crabs carry their fertilized eggs until they hatch as miniature replicas of the adult. Most marine crabs, though, release their offspring into the sea, where they develop through free-swimming larval stages.

As a crab larva ages, it gains a hard shell made of chitin, a natural polymer somewhat like plastic. To grow, the crab must periodically shed its rigid shell; this process is called molting.

Beneath a plume of bubbles, two crabs mate face to face. The male clasps the female with his pincers.

A crab comes of age

Female marine crabs carry their eggs for a short time after fertilization, then release the larvae into the ocean; there, for the next two months, the young crabs undergo drastic body changes.

The egg shown at near right (1) has been incubating for about 20 days; it has developed eyes and will soon hatch. As a just-released zoea (2), the minute larva floats atop the sea, molting up to seven times. After the last molt, it enters the megalops stage (3), taking on a crablike shape and a tough, chitinous coat. About two weeks later, another molt brings it to the "first" crab stage (4). After reaching the adult stage (5), the crab molts up to 20 more times in its three-year life span.

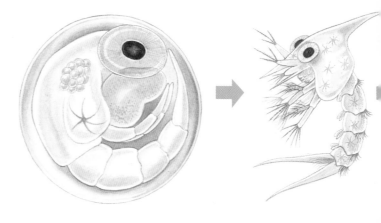

1) Development of eyes   2) Zoea larva

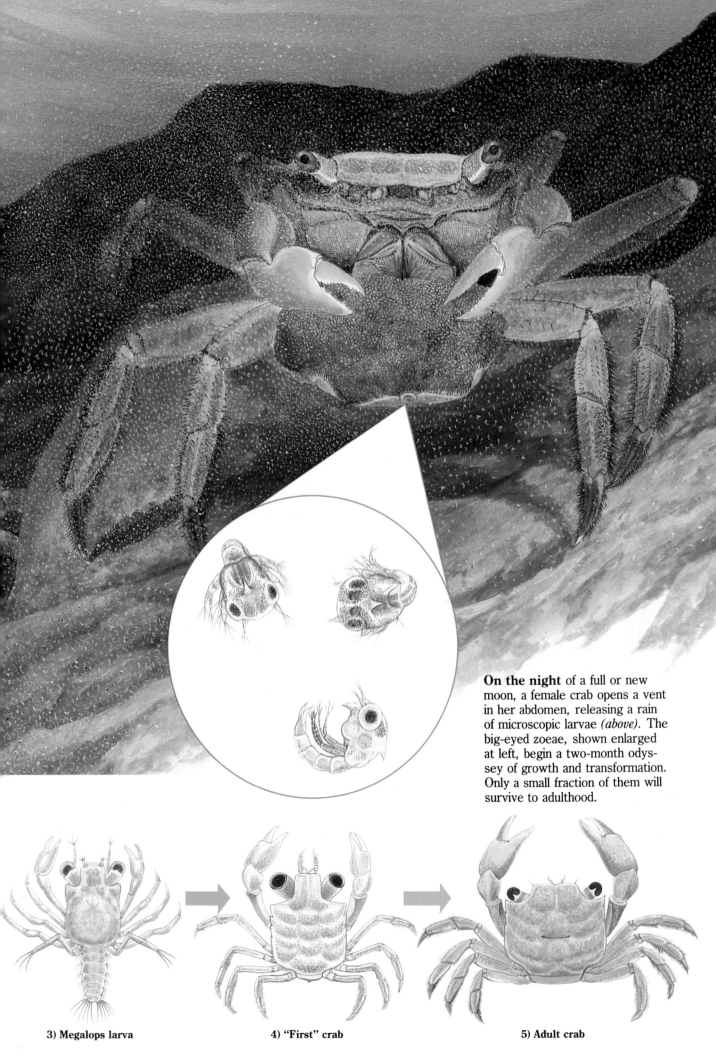

On the night of a full or new moon, a female crab opens a vent in her abdomen, releasing a rain of microscopic larvae *(above)*. The big-eyed zoeae, shown enlarged at left, begin a two-month odyssey of growth and transformation. Only a small fraction of them will survive to adulthood.

3) Megalops larva

4) "First" crab

5) Adult crab

How Do Mollusks Grow?

Free-floating abalone eggs develop rapidly after fertilization.

The eggs grow into larvae, which swim with hairlike cilia.

A metamorphosed larva takes on the shape of an adult.

The larvae settle to the sea bottom. They can now creep about.

Adults take in water and void wastes through shell holes.

The cilia grow longer, easing movement and feeding.

One slow mother

The giant mud snail of South America (*right*) lives in streams and rivers that remain at a fairly constant water level year round. Blessed with this stable environment, the mud snail runs little risk in laying fertilized eggs that hatch and develop outside the mother's body.

Other freshwater mollusks, however, are not so lucky. The Japanese mud snail, for example, has adapted to the sudden droughts of its home waters by practicing internal brooding: Eggs are fertilized inside the mother, where they hatch and mature into baby mollusks; the young are born fully formed.

Reproduction methods vary widely among mollusks, whose 100,000 species include such edible shellfish as clams, oysters, mussels, abalones, and snails. Some mollusks lay eggs. Others release larvae. Still others brood their young.

The single-shelled red abalone shown at left dumps clouds of sperm and eggs into the sea; fertilization takes place when the two gametes combine. Some sessile, or attached, bivalves—including the oysters shown below—also rely on this haphazard tactic.

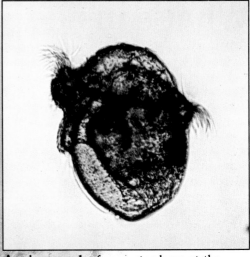

A micrograph of an oyster larva at the trochophore (free-swimming) stage reveals that the larva has sprouted tiny cilia and is beginning to build its shell.

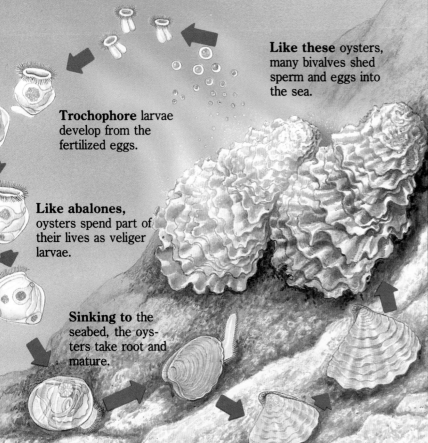

Like these oysters, many bivalves shed sperm and eggs into the sea.

Trochophore larvae develop from the fertilized eggs.

Like abalones, oysters spend part of their lives as veliger larvae.

Sinking to the seabed, the oysters take root and mature.

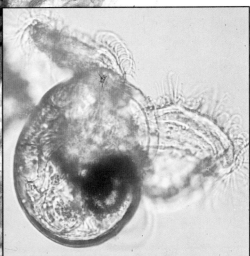

By the time the oyster larva reaches the veliger (intermediate) stage, its shell is well defined and its cilia make it self-propelling.

Armored eggs

Single-shelled mollusks—also known as univalves—generally engage in copulation, with the result that the eggs are fertilized within the body of the mother. Although the eggs are laid before they reach the larval stage, they are protected by a tough casing that increases their odds of survival. In species like the false trumpet, shown at right, the female glues the cases to a solid surface as she lays them. From the cases will emerge either veliger larvae or tiny mollusks that are already able to crawl.

A female false trumpet *(above)* lays her egg cases in a neat row, cementing each one in place. The eggs remain secured *(left)* until they hatch.

What's inside Sponges and Corals?

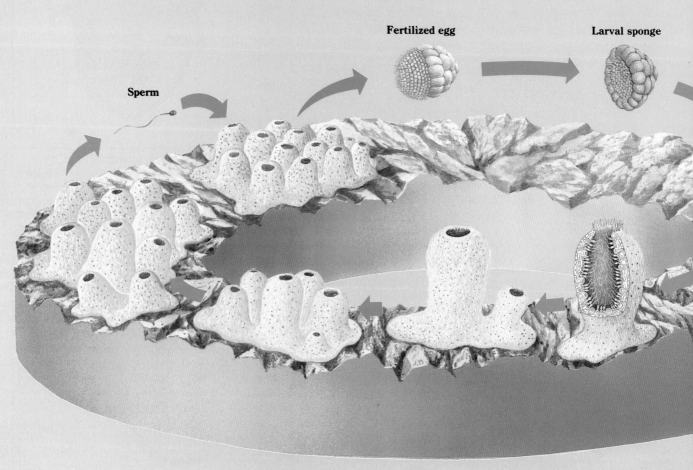

Sperm

Fertilized egg

Larval sponge

■ The life cycle of coral

A colony of coral usually spreads through budding. As shown at right, however, some corals can reproduce sexually. Sperm released from the gastrovascular cavity enters the mouth of a polyp and fertilizes an egg. The egg stays inside the polyp until it becomes a flagellated, or free-swimming, larva. The larva is ejected into the water, where it uses its cilia to swim or crawl away. When it finds a suitable spot, the larva attaches itself. The budding process then forms new polyps, which filter the seawater for nutrients. The colony proliferates until the cycle begins anew.

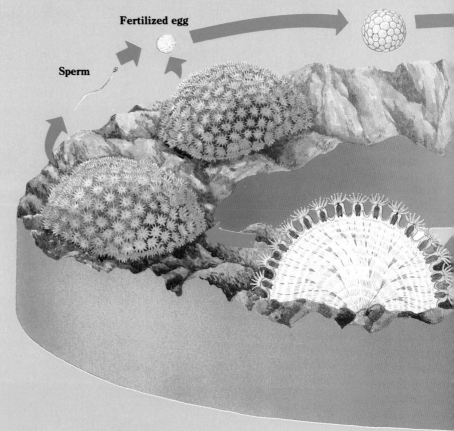

Fertilized egg

Sperm

Sponges and stony corals are not plants; they are primitive animals. Sponges, found in shallow water as well as deep, are made up of specialized cells, each performing its own chemical function. To reproduce, a sponge may release gemmules—clusters of cells that become new sponges—or it may bud, growing tiny offshoots.

Tropical and subtropical stony corals consist of two parts: colonies of starlike polyps, which are alive, and a skeletal structure of calcium carbonate that encloses the polyps. Like sponges, corals reproduce asexually—that is, they simply bud, or the polyps may split. But both creatures can also reproduce sexually, spewing eggs and sperm that yield free-floating larvae.

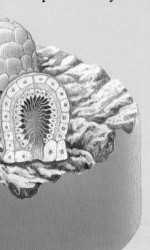

■ A sponge's life cycle

Depending on its environment, a sponge may bud, send out gemmules, or—as shown at left—release sperm in order to reproduce. Once a sperm cell combines with an egg, the fertilized egg is ejected, and the egg changes from generalized cells into specialized ones. This embryonic sponge then touches down, and its cells begin building a flexible or spiky framework, according to the species. Under favorable conditions, the sponge will grow and bud repeatedly.

Flagellated larva

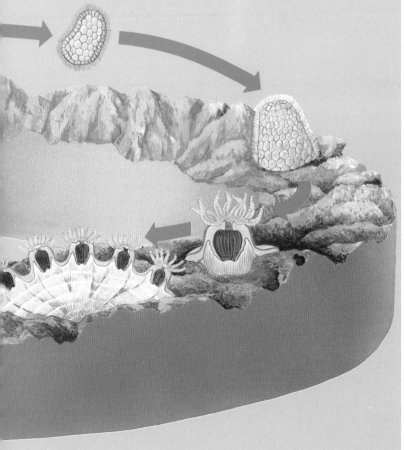

A tough case to crack

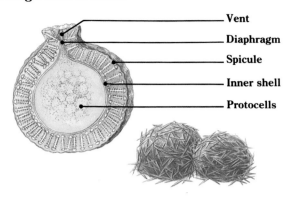

Vent
Diaphragm
Spicule
Inner shell
Protocells

Spiny spicules and a hard inner shell enable a gemmule—shown inside and out, above—to survive apart from its parent sponge. Gemmules are produced asexually when the water temperature plunges or the food supply dwindles. Later on, when conditions improve, the gemmule emits from its vent a cluster of primordial cells that grows into a full-fledged sponge.

How a sponge absorbs punishment

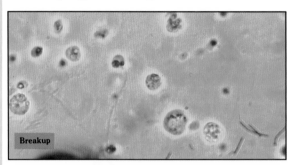

Breakup

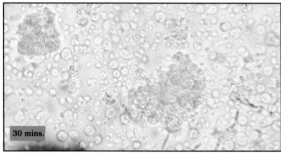

30 mins.

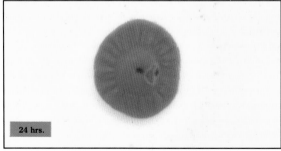

24 hrs.

The sequence of micrographs above reveals the speed with which a damaged sponge is able to repair itself. The sponge fragments *(top)* lack nerves and muscles, yet they begin to coalesce after just 30 minutes *(middle)*. Within 24 hours, the sponge has spontaneously reappeared *(bottom)*.

4
Desperately Seeking Sustenance

Perhaps nowhere in nature is the battle for survival tougher than in the water. Fish and aquatic mammals are well equipped to hunt and capture prey of every shape and size, so all but the largest animals living in water have a good chance of becoming another creature's dinner.

Marine predators use a variety of techniques to track and snare their victims. Blue whales *(pages 116-117)* and sardines simply open their mouths and swim through the water, catching

microscopic plankton in their baleen plates or gills. The deep-sea anglerfish is much more devious; a glowing lure on top of its head draws fish near, allowing the anglerfish to suck them into its mouth. Solo hunters—notably sharks and barracuda—succeed by the swiftness of their attack. Other creatures rely on cooperation; dolphins, for example, work in groups to drive large schools of fish into a small area, where they can be easily devoured.

Aquatic creatures hunt food in ways that are passive, aggressive, or deceptive. Sardines *(above, left)* swim with mouths agape, combing plankton from the water. Barracuda and sharks *(above)* bag their prey with blinding speed, brute force, and sharp teeth. A goosefish *(below)* gobbles up the smaller fish attracted to its lure.

Do Fish Have Favorite Foods?

Of the 25,000 fish species that inhabit the waters of the world, each has its own favorite food. In most cases, the shape and structure of a fish's mouth have evolved over eons to help it acquire that food. Fish that eat plankton, for example, have few if any teeth; they can expand their mouths like bellows to suck in and filter the food-filled water. Fish that feed on crabs and snails, by contrast, have large, molarlike teeth and powerful jaws that allow them to crush the creatures' shells. Sharks, which prey mostly on slower fish, have sharp teeth ideal for tearing chunks of flesh from their victims.

Triangular teeth set in muscular jaws help the great white shark rip its prey into bite-size chunks.

Needle-sharp teeth line the jaws of the barracuda, which snares its prey with hit-and-run strikes.

The skate catches bottom-feeding fish in its mouth, located on its underside. The skate's flat teeth crush the prey.

The moray eel's jaws open wide to capture large prey, such as this octopus. Caninelike teeth grasp slippery prey.

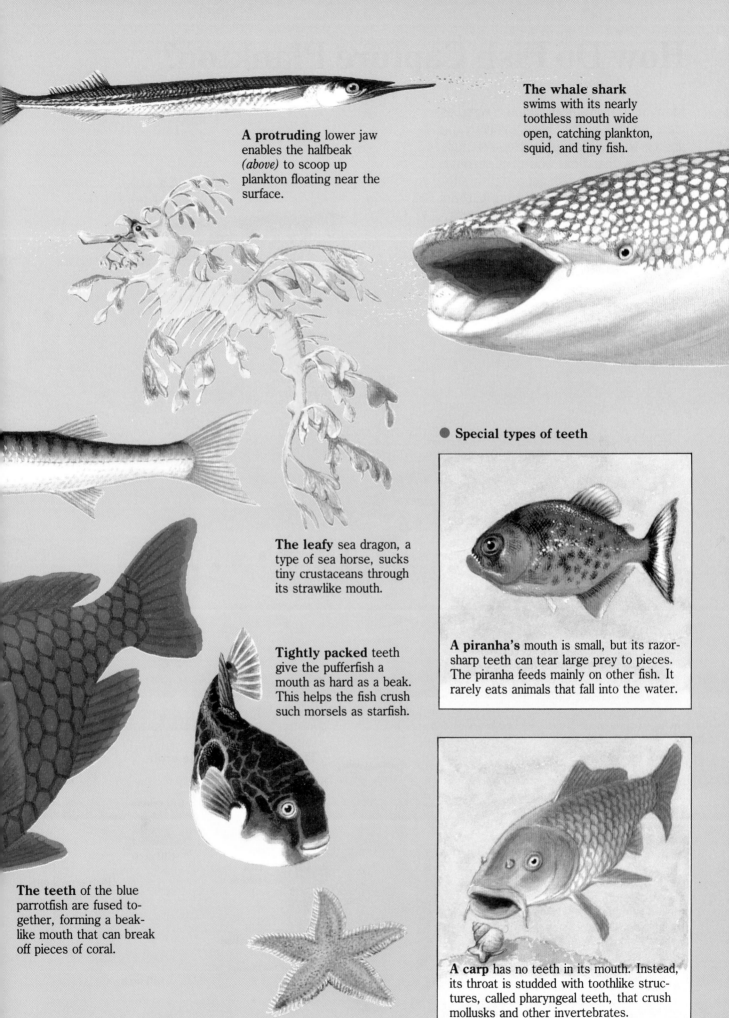

A protruding lower jaw enables the halfbeak (*above*) to scoop up plankton floating near the surface.

The whale shark swims with its nearly toothless mouth wide open, catching plankton, squid, and tiny fish.

The leafy sea dragon, a type of sea horse, sucks tiny crustaceans through its strawlike mouth.

Tightly packed teeth give the pufferfish a mouth as hard as a beak. This helps the fish crush such morsels as starfish.

The teeth of the blue parrotfish are fused together, forming a beaklike mouth that can break off pieces of coral.

● **Special types of teeth**

A piranha's mouth is small, but its razor-sharp teeth can tear large prey to pieces. The piranha feeds mainly on other fish. It rarely eats animals that fall into the water.

A carp has no teeth in its mouth. Instead, its throat is studded with toothlike structures, called pharyngeal teeth, that crush mollusks and other invertebrates.

How Do Fish Capture Plankton?

Many fish, from tiny sardines *(right)* to giant whale sharks, have evolved the ability to feed on plankton—minute creatures that float and drift through the ocean. These fish get a steady supply of food simply by opening their mouths as they swim through plankton-filled waters. Organisms are filtered from the water by bony gill rakers as the water passes over the fish's gills.

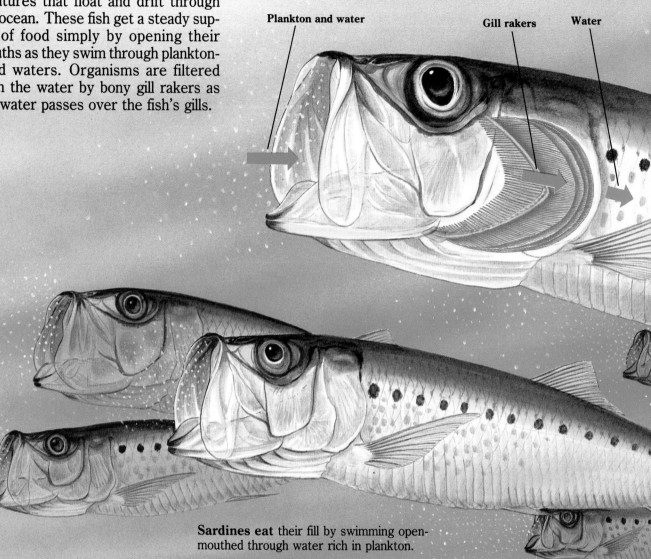

Plankton and water

Gill rakers

Water

Sardines eat their fill by swimming open-mouthed through water rich in plankton.

Thank heaven for little gills

Many fish have gill rakers, which keep food in the mouth but allow water to flow out. Made of bone or cartilage, the gill rakers grow from the gill arch. They resemble the teeth of a curved comb.

The size of the gill rakers depends on the food a fish eats. Plankton eaters have long, tightly packed gill rakers *(right, top)* that strain the microorganisms from the water. Plankton lodge in the gill rakers and are washed down the throat by water. Fish eaters, by contrast, have no need to trap plankton; their gill rakers are small and far apart *(right, bottom)*.

The right gills for the grub

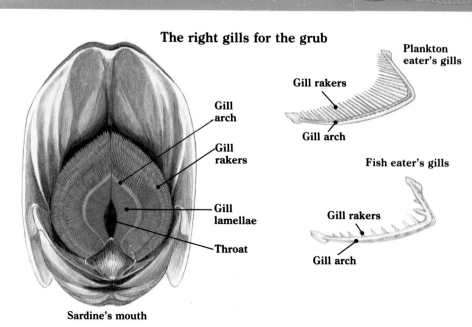

Gill arch

Gill rakers

Gill lamellae

Throat

Sardine's mouth

Plankton eater's gills

Gill rakers

Gill arch

Fish eater's gills

Gill rakers

Gill arch

Gentle giants

Some of the largest aquatic animals feed solely on plankton and the tiny crustaceans known as krill. One such so-called filter feeder is the blue whale, the largest creature on Earth. The blue whale catches up to 8 tons of krill per day on combed plates collectively called baleen. Giant sharks, such as the 60-foot whale shark and the 35-foot basking shark, trap plankton on gill rakers. Many other seemingly fearsome fish—including those shown at right—subsist on an all-plankton diet.

The paddlefish (*above*) was named for its flattened, spatula-shaped upper jaw. It lives in the Mississippi River basin, where it may grow to nearly 6 feet.

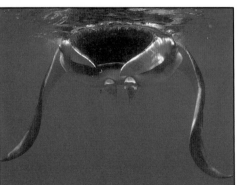

The body of a manta ray may measure up to 20 feet across and weigh as much as 1½ tons. The ray uses its front fins to funnel plankton into its mouth.

What Is an Archerfish?

Most predatory fish attack creatures that live in the water. But the archerfish and its relatives, which live in coastal and fresh waters from India to Australia, have extended their hunt to the air space above their aquatic home. These fish can shoot a jet of water into the air, knocking an insect off a plant or out of the sky. Stunned by the blow, the insect falls onto the water's surface, where the archerfish makes a quick meal of it.

If the prey is less than 3 feet above the water, one shot usually brings it down. But if the insect is 3 to 10 feet above the water, the archerfish may need to hit it several times. When the quarry lingers just over the surface, the archerfish may leap from the water and snatch it in its jaws.

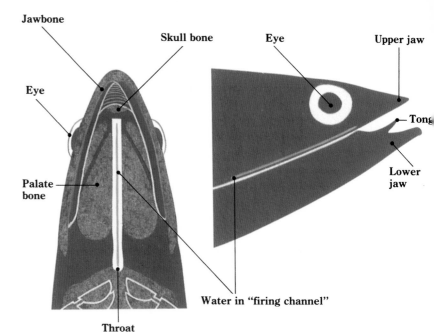

When the archerfish closes its mouth, its hard tongue and upper jaw form a firing channel. The archerfish pumps water through this channel by closing its gill covers.

Jawbone
Skull bone
Eye
Upper jaw
Eye
Tong
Palate bone
Lower jaw
Water in "firing channel"
Throat

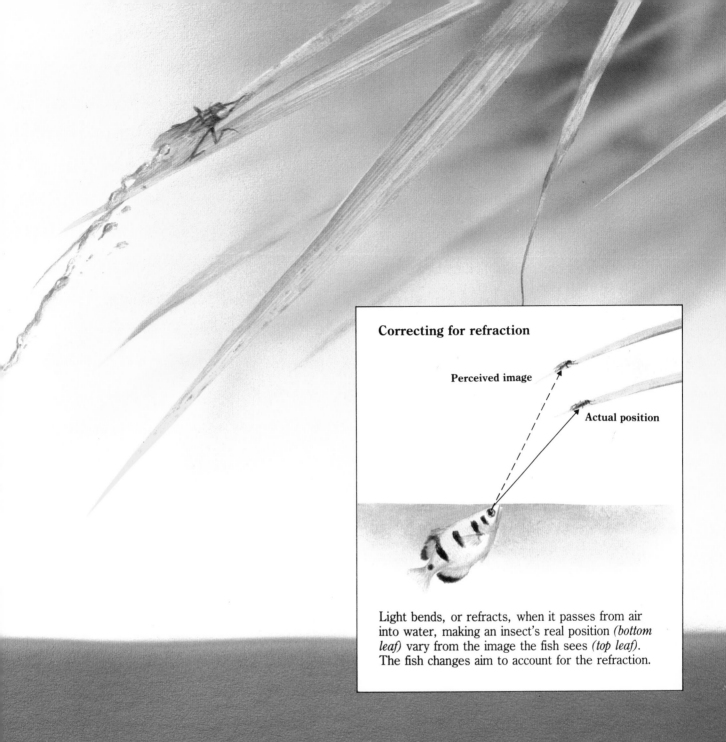

Correcting for refraction

Perceived image

Actual position

Light bends, or refracts, when it passes from air into water, making an insect's real position *(bottom leaf)* vary from the image the fish sees *(top leaf)*. The fish changes aim to account for the refraction.

Catching prey

The archerfish is a resourceful hunter. In addition to its sharp-shooting skills, the fish is a powerful jumper; as shown at right, it can leap from the water to grab an unsuspecting insect off a leaf or branch overhanging the water. The archerfish will also snap up insects or small creatures that land on the water's surface. When no individual targets can be found, the archerfish rises to the surface and squirts a stream of water droplets into a swarm of flying insects *(far right)*, knocking at least one to the surface.

An archerfish leaps to snatch prey from a leaf.

A flying insect is hit by an archerfish's random shot.

Do Fish Go Fishing?

Some aquatic creatures use trickery to draw their prey within striking distance. Anglerfish, for example, boast an appendage that resembles a fishing pole and lure; this structure is in fact the first spiny ray of the anglerfish's dorsal fin, which grows on its back. When a smaller fish approaches, the anglerfish gets its attention by raising the lure above its mouth and waving it back and forth. As soon as the smaller fish nibbles at the lure—it may even bite off a small piece—the anglerfish opens its mouth and gobbles up the fish. The lure may be damaged in the fray, but it quickly regenerates.

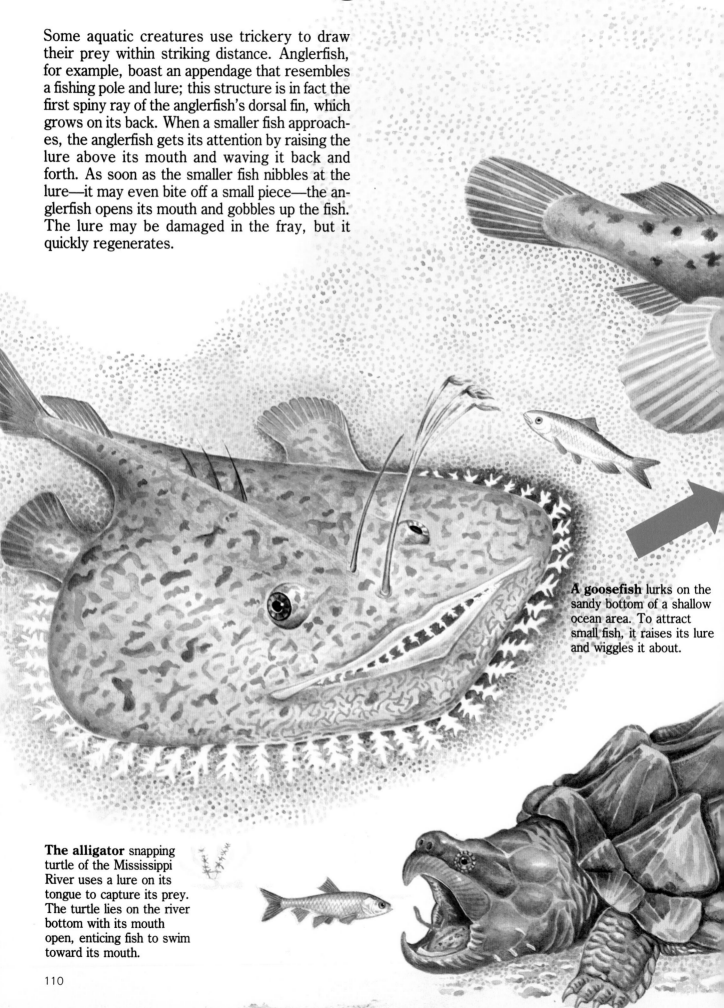

A goosefish lurks on the sandy bottom of a shallow ocean area. To attract small fish, it raises its lure and wiggles it about.

The alligator snapping turtle of the Mississippi River uses a lure on its tongue to capture its prey. The turtle lies on the river bottom with its mouth open, enticing fish to swim toward its mouth.

Mistaking the lure for food, a smaller fish swims toward it. The goosefish suddenly opens its mouth and swallows the prey.

The frogfish *(left)* draws fish near with its lure; it then opens its mouth rapidly, creating a current that sucks the prey inside.

▲ **Light-producing** bacteria live in the lures of deepwater female anglerfish. The light attracts both prey and potential mates.

How Do Deep-Sea Fish Catch Prey?

About 12 percent of all fish live in the perpetual dark of the deep ocean, 3,000 feet or more below the nutrient-rich waters of the surface. Food is scarce at these depths, so the creatures who live here have become skilled at hunting without wasting energy. Some can float suspended in the water without moving their fins. Others spend their lives on the ocean bottom, using their fins as legs to crawl from one spot to another. Most deep-sea fish have small eyes, which may not function well. They have a keen sense of smell, however, and can hear or feel the slightest vibration—a signal that the next meal may be nearby.

1 **Three small fish,** accustomed to eating bioluminescent creatures, investigate the glowing lure of a deep-sea anglerfish. The anglerfish will quickly snap them up.

2 **The Atlantic snipe eel** floats vertically just above the seabed. Its slender jaws curve outward to snare the antennae of shrimp and other crustaceans that happen to swim by.

3 **The mouth of the gulper** is as long as its head, enabling the eel to swallow almost any creature its own size that swims within range. Its whiplike tail propels it along the ocean floor.

4 **Many denizens of the deep**—including the one eating a squid above—have large mouths and expandable stomachs, permitting them to make the most of infrequent meals. The fish in this realm also eat dead creatures that drift down from the waters above.

5 **This bottom-dwelling spiderfish** rests on its pelvic and tail fins, facing into the ocean current. It then uses its pectoral fins as a fishing net, snaring small crustaceans as they float by.

How Do Dolphins and Whales Hunt?

Although dolphins look like fish, they are actually mammals. Dolphins and their relatives the whales are both descended from a catlike predator that returned to the sea some 55 million years ago. Like many other mammals, dolphins and whales are social creatures. They hunt in packs, and they even have their own languages. Dolphins and whales find their meals using echolocation: They emit sound pulses that reflect off schools of fish, revealing their precise position.

Killer whales, or orcas, use the underwater terrain to advantage when hunting. Acting as one, the orcas herd a school of fish against a cliff, then eat the entire group.

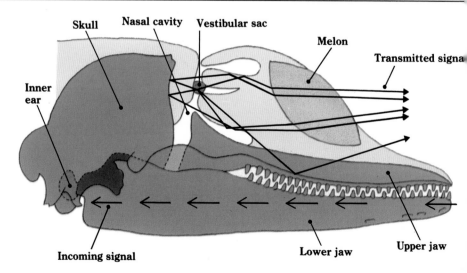

Sending and receiving sounds

Unlike other mammals, which make sounds with their vocal cords, dolphins and whales produce their characteristic clicks and whistles in their skull cavities. Air enters the nasal cavity through the blowhole; as the air moves back and forth inside this cavity, it vibrates pieces of tissue called nasal plugs. The melon—a pad of fat found only in aquatic mammals—acts as a lens, focusing the sound waves so they emerge from the skull as a tight beam. Incoming signals travel through the lower jaw before striking the inner ear at the base of the animal's skull.

Skull Nasal cavity Vestibular sac

Melon

Transmitted signal

Inner ear

Incoming signal

Lower jaw

Upper jaw

Spinner dolphins

Bridled dolphins

Bridled dolphins, found off Argentina, hunt in a long, evenly spaced line. Each dolphin swims about 35 feet from the next.

Bridled dolphins and spinner dolphins often hunt together. In the morning, the bridled dolphins hunt while the spinners watch for predators such as sharks and orcas. In the evening, the two switch jobs.

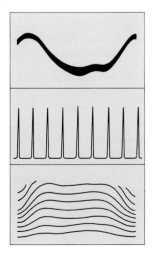

Whistles and clicks

Dolphins and whales produce three distinct types of noise. To communicate with one another, they use a high-pitched whistle *(left, top)*. When searching for food, dolphins emit a rapid and rhythmic click-click-click *(middle)* that reflects off any object it hits. Like a submarine using sonar, the dolphin hears the returning clicks and uses them to zero in on its prey. Stratified sound *(bottom)* is a blend of whistles and clicks.

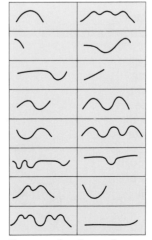

Conversational sounds of bottle-nosed dolphins

Underwater conversations

Like all effective communicators, dolphins and whales have rich vocabularies *(left)*. The members of a school use the same collection of sounds when they communicate. This collection of sounds is also used by other schools of the same species and similar social structure, suggesting a common language. A school of the same species but with a different social structure, however, uses a different set of sounds, which may be a dialect of the main language.

What Do Blue Whales Eat?

Measuring 100 feet from tip to tail and weighing up to 150 tons, the blue whale is the world's biggest animal. It lives on a diet of tiny organisms—shrimplike crustaceans known as krill—that rarely grow more than 2½ inches long.

A blue whale may eat up to 8 tons of krill in just one day. Such feasting requires two unusual features. The first is the baleen—ridged, bony plates in the upper jaw that trap krill but allow water to pass. The second is a set of approximately 60 parallel grooves running from the whale's lower jaw to its navel. The grooves enable the whale to expand its mouth, taking in 1,700 gallons of water and krill in a single gulp.

An adult blue whale like the one below uses its baleen—comblike plates attached to its upper jaw—to trap some 800 million krill every day.

An efficient filter

The blue whale's baleen plates, which resemble a thick but stubby brush, stretch the length of its upper jaw *(right, top)*. The plates function much like the gill rakers of plankton-eating fish. After closing its mouth around a cloud of krill *(bottom left)*, the whale pushes its muscular tongue against the plates; at the same time, it contracts its stretched mouth *(bottom right)*. These two actions force water through the baleen plates and out the sides of the whale's mouth. The whale then swallows the krill that have been strained out by the baleen.

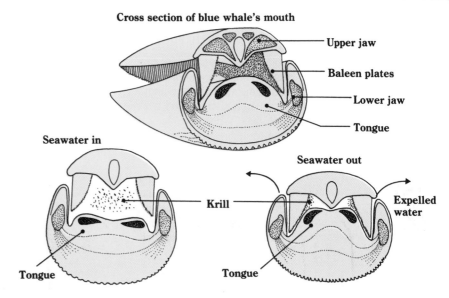

Cross section of blue whale's mouth

Upper jaw

Baleen plates

Lower jaw

Tongue

Seawater in

Seawater out

Krill

Expelled water

Tongue

Tongue

A whale of a creature

Not only is the blue whale the largest animal alive today, it is also the largest creature that ever lived. The dinosaur *Brachiosaurus,* for example, was 80 feet long and weighed 80 tons; the African elephant weighs a mere 7½ tons.

0 (ft)	33	65	100

Fatty insulation

Because krill abound in cold waters, blue whales spend much of their lives in the oceans of the Arctic and Antarctic. The whales manage to survive this icy realm thanks to a thick layer of blubber, or fat, just beneath their skin *(far right)*. Fat is a poor conductor of heat, so the blubber acts like a heavy fur coat. The one spot vulnerable to heat loss is the whale's mouth; no blubber protects the baleen plates. The arteries in the plates are therefore surrounded by veins, which insulate the arteries from the cold water.

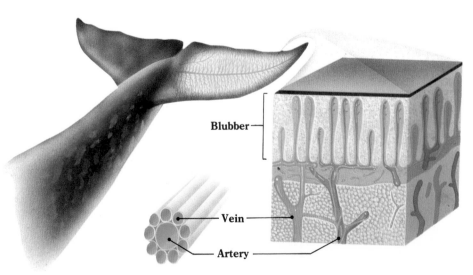

Blubber

Vein

Artery

Can Shellfish Attack Fish?

The snail-like creatures known as cone shells are much too slow to chase their prey. Instead, they stab poison-filled teeth at fish, worms, and snails that pass within their reach. The potent toxin paralyzes prey within seconds, giving the cone shell plenty of time to catch up to—and consume—its helpless victim. The poisons of some cone-shell species—notably those that live in the southern Pacific Ocean—are so toxic they can kill a human being in four hours.

Like snails, cone shells creep through their environment on a single muscular foot. A pair of eyes and two antennae help the cone shell spot its prey. Its breathing organ is called a siphon. The rest of the animal's body is hidden within its long, protective shell.

Extending its flexible proboscis, a deadly cone shell fires a radula tooth filled with venom into a passing goby.

Inside a poisonous slowpoke

The cone shell stuns its prey with a special barb, called a radula tooth, that acts as a poison dart. Preparing to attack, the cone shell extends its long, highly maneuverable proboscis. This causes a single radula tooth to be released from the long arm and "loaded" into the proboscis; along the way, the tooth fills with a lethal nerve toxin produced in the poison sac. When the barbed end of the tooth is thrust into the prey, the toxin flows into the creature, rapidly paralyzing it. In cone shells that attack snails, the radula tooth detaches from the proboscis upon entering the victim. In cone shells that feed on fish, by contrast, the tooth remains attached to a ligament, like a harpoon.

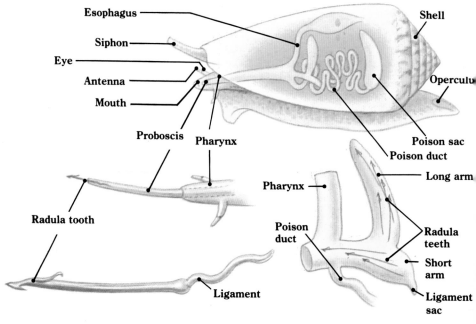

Esophagus

Siphon

Eye

Antenna

Mouth

Proboscis

Pharynx

Radula tooth

Ligament

Shell

Operculum

Poison sac

Poison duct

Long arm

Pharynx

Poison duct

Radula teeth

Short arm

Ligament sac

With its prey paralyzed by a powerful nerve toxin, the cone shell moves in to consume the helpless goby.

The cone shell opens its mouth wide enough to swallow the victim whole.

Tropical predators

Cone shells flourish in the tropical coral reefs of the Indian, South Pacific, and western Atlantic oceans. Some cone shells nestle in the sandy ocean bottom, waiting for a fish to pass overhead. Others lie in wait for unsuspecting snails *(right)* or for the lugworms *(far right)* that slither along the ocean floor.

The radula tooth of a fish-eating cone shell remains attached to the proboscis by a ligament after an attack, enabling the cone shell to reel in its meal. The radula tooth of a snail-eating cone shell, however, detaches upon impact, forcing the cone shell to crawl after its victim and dislodge it from the shell.

Having poisoned a snail, a cone shell proceeds to remove the dead creature from its shell.

Like a strand of undersea spaghetti, a flat lugworm disappears inside the mouth of a cone shell.

5

The Art of Self-Defense

In the crowded aquatic food chain, many a predator is prey for another species. To avoid being eaten and to survive long enough to bear young, underwater creatures have developed special ways of protecting themselves.

A flying fish, for example, eludes its enemies by going where they cannot follow: into the air. Yet even this tactic is not foolproof; once airborne, the flying fish may be gobbled in midflight by an albatross or a sea gull.

Other species hide themselves. The decorator crab blends into its surroundings by draping itself

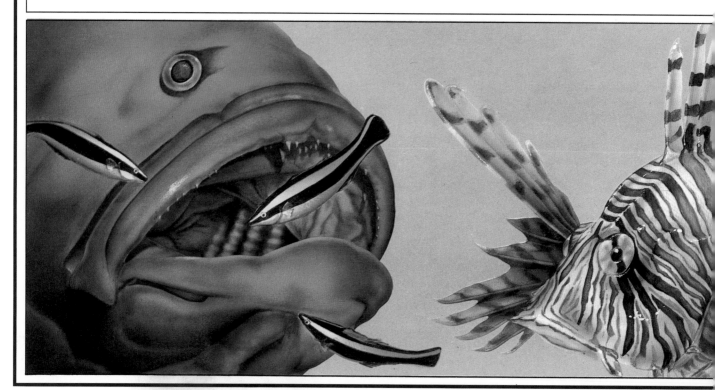

with bits of seaweed. Flounder have no need for such props; they simply change the color of their bodies to match that of the seabed.

Some creatures use weapons to repel their enemies. Stinging spines, each capable of delivering a jolt of poison, protect the lionfish as it swims. When danger threatens a porcupinefish, the creature turns into a prickly globe too big to be swallowed by most fish. Squid and octopuses thwart pursuers by squirting ink into the water.

And when a fish has no weapons of its own, it may pair up with another creature that is better armed. Clownfish like the two on the cover of this book take haven among the stinging tentacles of a sea anemone, while cleaner wrasses stick close to larger fish that provide food and scare off potential predators.

Skimming the surface or gliding over it in graceful arcs, a school of flying fish (above) flee an enemy. Below, from left to right, three cleaner wrasses swim near their host, a lionfish harbors poison spines in its dorsal fin, and a porcupinefish puffs up in self-defense.

What Are Poison Fish?

Just as some snakes protect themselves by injecting poison through their fangs, some fish defend themselves with poisonous spines located on their body or fins.

A stingray can deliver a dose of poison from a spine at the base of its tail. Colorful lionfish and homely lumpfish both have poisonous spines on their backs. And rabbitfish bristle with toxic spikes on their dorsal, pectoral, and tail fins.

In some species, the poison is contained in a needlelike tip, which is used to pierce the victim's flesh. In others, such as the stingray, toxins are released from skin surrounding the spine. The poisons vary in strength, but only a few of them may be fatal to humans.

Catfish like this one live in rivers. Some are armed with poisonous spines in their dorsal and pectoral fins.

Lionfish

Probably named for its mane-like pectoral fins, the lionfish swims without fear near the ocean bottom, protected by poisonous spines along its dorsal fin.

122

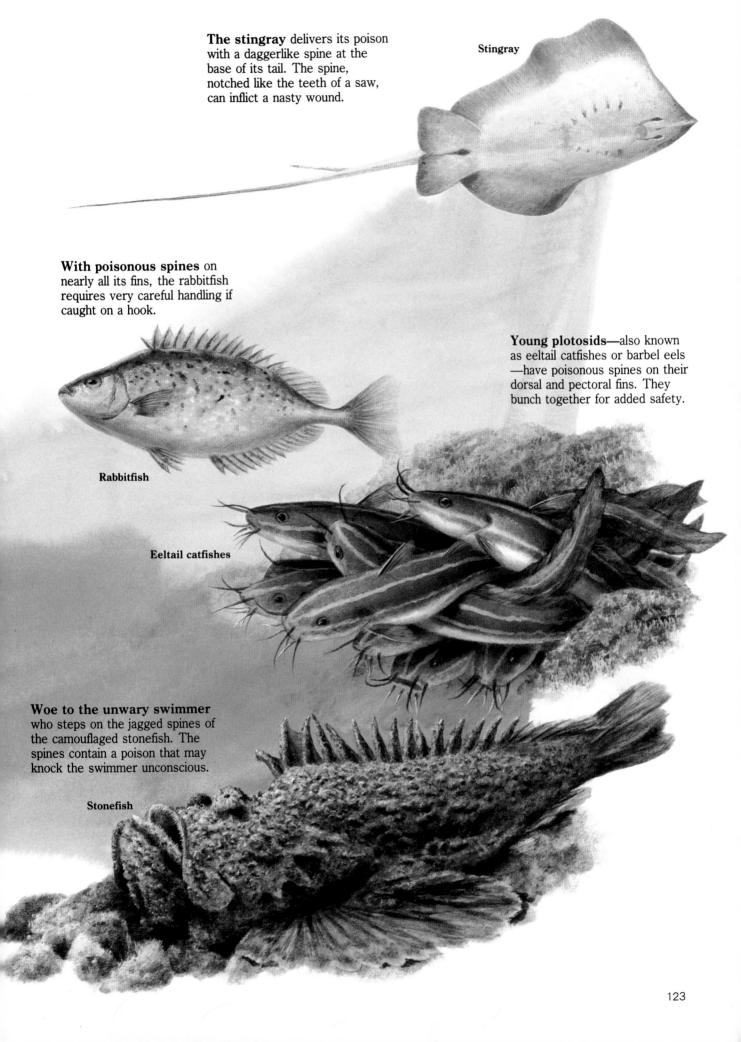

The stingray delivers its poison with a daggerlike spine at the base of its tail. The spine, notched like the teeth of a saw, can inflict a nasty wound.

Stingray

With poisonous spines on nearly all its fins, the rabbitfish requires very careful handling if caught on a hook.

Young plotosids—also known as eeltail catfishes or barbel eels —have poisonous spines on their dorsal and pectoral fins. They bunch together for added safety.

Rabbitfish

Eeltail catfishes

Woe to the unwary swimmer who steps on the jagged spines of the camouflaged stonefish. The spines contain a poison that may knock the swimmer unconscious.

Stonefish

What Is a Porcupinefish?

Sensing danger, a porcupinefish erects its spines, which normally lie nearly flat on its body. On a full-grown fish, each spine may be at least 2 inches long.

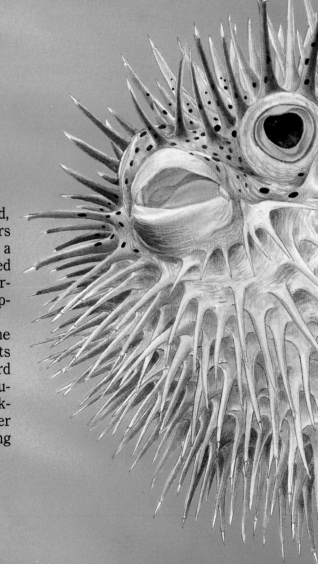

Like a hedgehog raising its quills to defend itself on land, the porcupinefish erects sharp spines to ward off predators in the sea. When an enemy approaches, the fish inflates a stomach sac with water or air, transforming its streamlined body into a bristly balloon that may be three times its normal size. Confronted by this prickly and unappetizing apparition, most wise predators keep their distance.

The porcupinefish lives in tropical waters around the world. Its favorite habitat is the coral reef, where it hunts chiefly at night, using its beaklike mouth to crush the hard shells of mollusks and crustaceans. If its prey proves elusive, this master of defense goes on the offensive; puckering its lips, the porcupinefish shoots a stream of water into the sandy seafloor, dislodging the clam or crab trying to hide there.

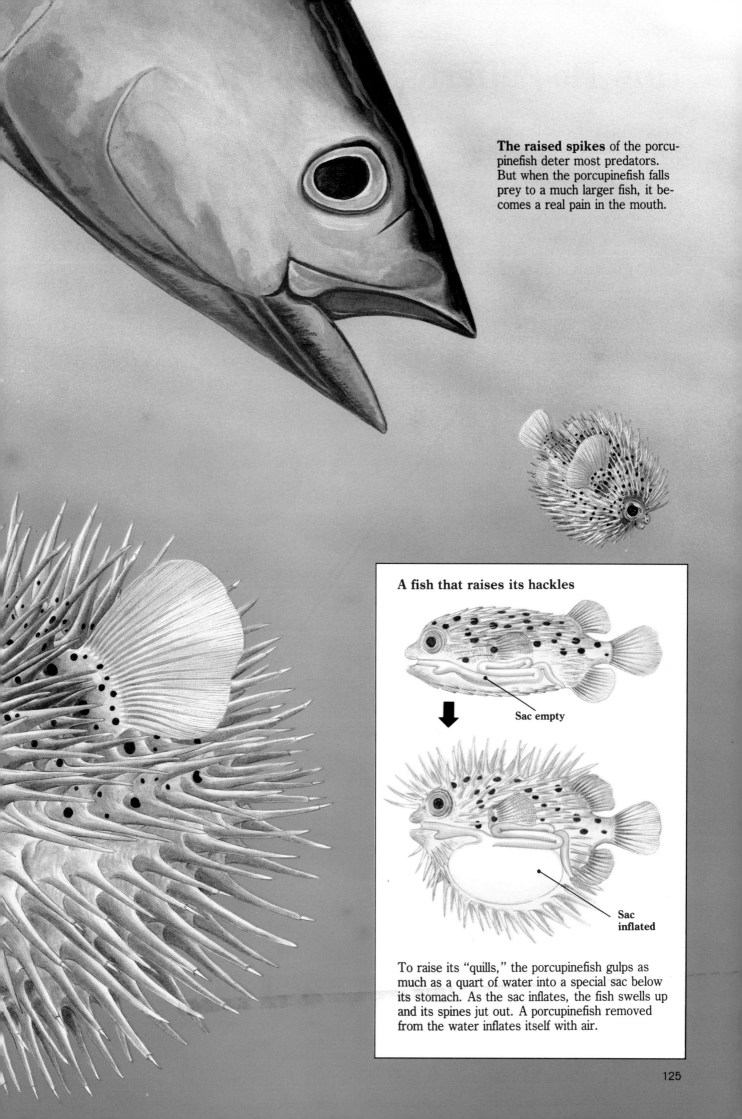

The raised spikes of the porcupinefish deter most predators. But when the porcupinefish falls prey to a much larger fish, it becomes a real pain in the mouth.

A fish that raises its hackles

Sac empty

Sac inflated

To raise its "quills," the porcupinefish gulps as much as a quart of water into a special sac below its stomach. As the sac inflates, the fish swells up and its spines jut out. A porcupinefish removed from the water inflates itself with air.

How Do Pufferfish Defend Themselves?

A pufferfish, as its name suggests, can puff up its body. It uses this tactic to scare off enemies. Many species of pufferfish boast a second line of defense—poison. Some of them may release their poison into the water to deter a pursuer, while others are poisonous only if eaten.

Oddly, a pufferfish may not make its own poison. Some scientists believe the poison is produced by bacteria or other microorganisms in the water and muddy seafloor of the tropical coastal areas where most pufferfish live. Working its way up the food chain—through lugworms, starfish, clams, snails, and other creatures *(right)*—the poison accumulates in the internal organs of pufferfish. Here the toxin may reach a concentration that is 150,000 times more potent than curare, the South American arrow poison.

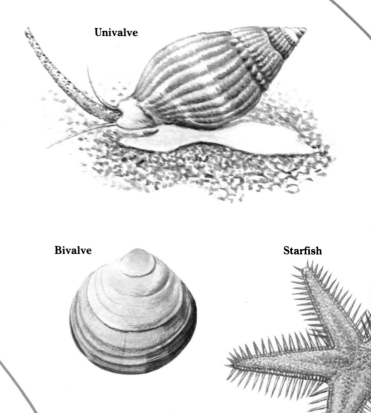

Univalve

Bivalve

Starfish

Bacteria

Lugworm

Deadly enemies

A relative of the pufferfish, the trunkfish, stores a similar poison in its skin. When the trunkfish senses danger, it releases the toxin into the surrounding water. In the open ocean, the poison wards off nearby enemies without harming the trunkfish. When a trunkfish is confined in an aquarium, however, its poison can kill other fish—and even the trunkfish itself. Soapfish secrete a different kind of poison from their skin when threatened.

Trunkfish

Yellow emperor (soapfish)

Golden-striped grouper (soapfish)

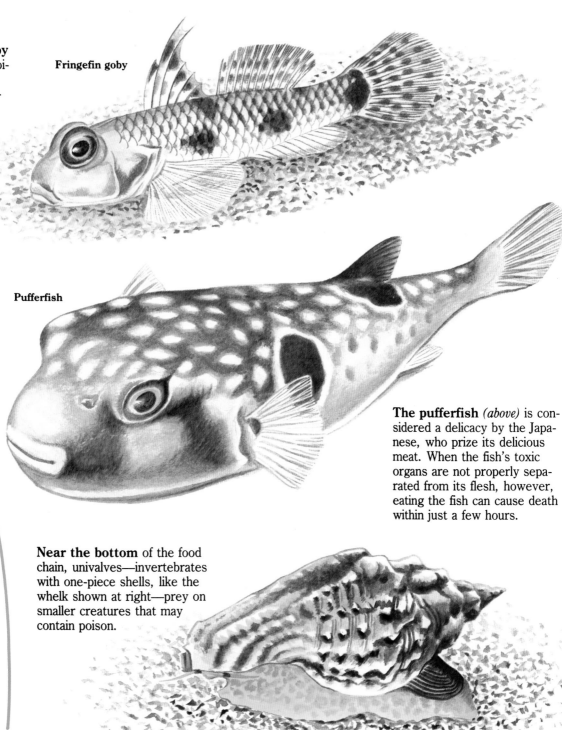

The fringefin goby carries the same poison found in many pufferfish. This poison was once used to kill rats in fields.

Fringefin goby

Pufferfish

The pufferfish *(above)* is considered a delicacy by the Japanese, who prize its delicious meat. When the fish's toxic organs are not properly separated from its flesh, however, eating the fish can cause death within just a few hours.

Near the bottom of the food chain, univalves—invertebrates with one-piece shells, like the whelk shown at right—prey on smaller creatures that may contain poison.

Poisonous fugu

Tetrodotoxin—the poison carried by many species of pufferfish, including those at right—is highly poisonous to humans. But aquaculturists have attempted to breed pufferfish whose bodies are free of the toxin. Such a strain of pufferfish would be safe to eat.

The first step is to create a fish nursery devoid of poison-producing microorganisms, so no poisons can enter the food chain. Next, toxin-free puffers mate. Over time, this strategy may produce nonpoisonous puffers.

Panther puffer

Genuine puffer

127

Do Flying Fish Really Fly?

Flying fish have a clever way of escaping pursuers: They take to the air, leaving their enemies behind. Although the fish cannot flap their fins, they can use them to glide over the water's surface for up to 215 yards at a time.

A number of special adaptations make this possible. The enlarged lower half of the tail fin, which beats the surface of the water up to 50 times a second, provides the speed needed for takeoff. Another crucial feature is greatly enlarged pectoral fins, which act like wings. Air moving across these fins raises the fish above the water, keeping it briefly airborne. The pelvic fins provide additional lift during flight.

Rapidly sculling its tail, the fish gains speed as it moves up toward the surface. Upon reaching 40 miles per hour, it is ready to take off.

To become airborne, the flying fish beats the surface of the water with the bottom lobe of its tail fin, providing thrust, and extends its pectoral fins, providing lift. As the fish clears the water, it spreads its pelvic fins, lifting its tail fin clear of the surface.

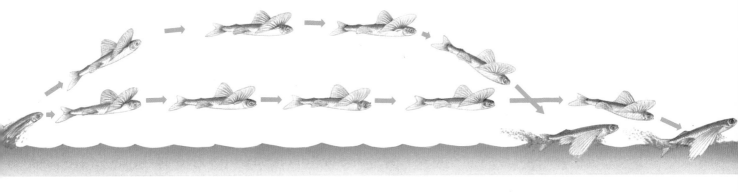

The fish that soared

After taking off at 40 miles per hour, the flying fish glides along just inches above the water for about 10 seconds. Gravity and wind resistance slow it to about 20 miles per hour, causing the fish to reenter the water. If danger still threatens at the end of its flight, the fish rapidly sculls its tail fin again to start another glide.

Long, stiff supports called fin rays run through the pectoral fins, making the fins strong enough to hold the fish aloft.

Partners in flight

Flying fish and gliders rely on similar features to fly. The pectoral fins of the fish and the wings of the glider are both rigid during flight. In addition, both the fins and the wings are tapered at the tips. This directs the airflow toward the junction between the fins (or wings) and the body, taking some of the strain off the fins and making flight easier. And just as gliders are made from lightweight materials, the fish have unusually small stomachs and intestines, reducing their flying weight.

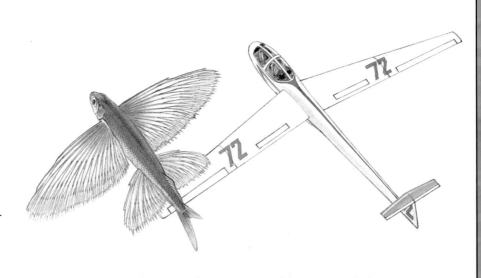

Why Are Flounder Flat?

Flounder—remarkable for their broad, flat bodies—live in shallow areas of all the world's oceans. Except when they are chasing prey, flounder spend their time lying motionless on the ocean floor. Their flat shape helps the flounder blend into the seabed, keeping them hidden from predators. For added camouflage, they change their color and markings to match those of the background.

Adult flounder appear flat because they spend their entire lives lying on one side. Much like other fish, flounder start their lives upright. A few days into their development, however, they gradually begin to tilt sideways. The eye on one side slowly migrates around the fish's body, moving into position on what will become the fish's top side. While this is taking place, the bottom side grows steadily lighter in color, and the upper side grows darker.

There are both left-eyed and right-eyed flounder. The side on which the fish lies is characteristic for each species.

1. Some flounder may lay up to three million eggs. The eggs soon get separated from one another; most float to the surface.

2. After developing for several days, the eggs hatch and larvae emerge. Lacking fully formed fins, the larvae drift with the currents and settle to the bottom.

3. As the larvae begin to develop fins, they become fry that vigorously swim about. At this stage, their body structure differs little from that of many other fish.

4. Several weeks into the flounder's development, the fins are well developed, but the eyes remain on opposite sides of the fish's head.

5. As the flounder's metamorphosis into a flatfish begins, one eye—in this case, the right one—migrates to the top of its head, directly above the backbone.

6. The right eye comes to rest on the left side of the body. The fish takes up residence on the ocean bottom, where it swims by undulating its entire body.

7. The flounder completes its transformation in a few months. Both eyes are now on top of its body, and pigment cells in its skin are the color of its surroundings.

8. As an adult, the flounder actively pursues such sea-bed creatures as crustaceans, smaller fish, squid, mollusks, worms, and other bottom-dwellers.

9. When not pursuing its next meal, a fully grown flounder lies still on the seabed or buries itself in the mud or sand, with only its eyes protruding.

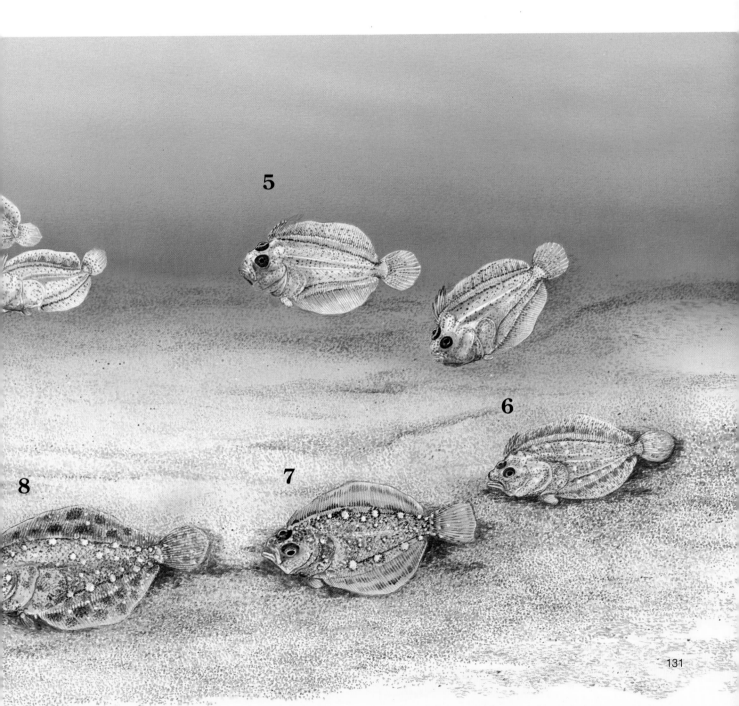

Why Do Angelfish Have Stripes?

An adult angelfish shares territory with no grown fish but its mate.

Distinctive markings on a young fish protect it.

Signs of a shared ancestor

The four angelfish species shown at right have nearly identical body markings—vertical blue and white stripes—when young. The prevalence of this pattern suggests that all four species may have had a common ancestor. The adult markings, by contrast, differ from one species to the next. This variance may be useful in keeping the adults of one species from mating with those of another species.

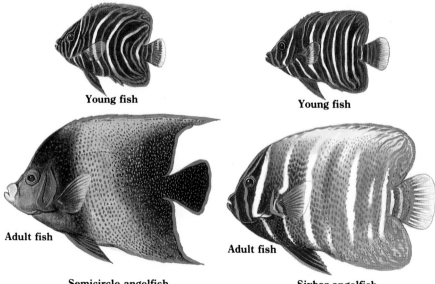

Young fish

Young fish

Adult fish

Adult fish

Semicircle angelfish

Sixbar angelfish

An adult angelfish, distinguished by its bright colors, guards its territory by attacking any other full-grown angelfish that ventures inside it. Young angelfish that enter the territory of an adult, however, are left undisturbed.

The reason lies in the blue and white vertical stripes that are worn by many juvenile angelfish, no matter what their species. This pattern marks an angelfish as immature—that is, not yet able to challenge an adult's territory. Until its blue and white stripes begin to change to the gaudy colors of an adult, the youngster will remain safe from attack.

As its marks change, the fish loses its protection.

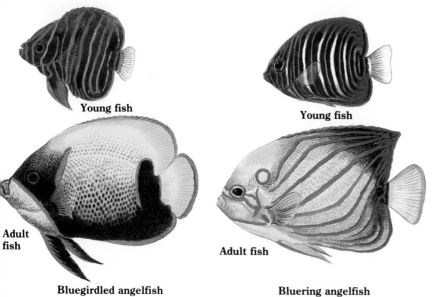

Young fish

Young fish

Adult fish

Adult fish

Bluegirdled angelfish

Bluering angelfish

A multicolored maturation

Rather than changing color when it reaches a certain age or size, the yellowface angelfish changes its markings according to its degree of interaction with other fish. Yellowfaces raised in a tank without adults, for example, begin to take on adult markings at an early age. Young raised in a tank containing adults, however, keep their juvenile markings even after growing quite large. The high risk of attack in the adult tank prompts the young fish to keep their protective juvenile markings as long as possible.

Very young fish have blue and white stripes on a black background.

As the black lightens, yellow and green begin to tinge the scales.

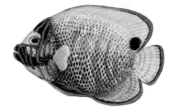

The stripes disappear everywhere but on the head.

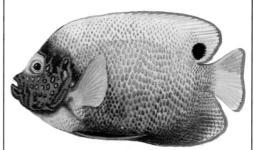

Speckles emerge as the dominant motif of the adult's markings.

What Are "Cleaner Fish"?

The cleaner fish, a kind of wrasse, lives near coral reefs in the Indian and Pacific oceans. Although it is only about 4 inches long, the cleaner fish moves fearlessly among larger, more aggressive fish. When approached by a grouper, for example, the cleaner fish begins to prod and peck at the normally dangerous predator; it may then swim safely inside the grouper's mouth, where it eats parasites and other harmful organisms. All sorts of fish, from snappers to sharks, seek out schools of cleaner wrasses to have parasites removed from their fins, bodies, gills, and mouths.

The distinctive head-down dance of the cleaner wrasse advertises its cleaning services to other fish.

Masquerading as cleaners

All three of the fish at right bear markings that make them look like cleaner fish. In reality, however, the top two are saber-toothed blennies. The impostors differ from true cleaner fish in three ways: Their mouths point down, their dorsal fins are longer, and their tail fins are more rounded. The saber-toothed blennies benefit from this similarity by executing a cruel hoax *(opposite page)* on larger fish that approach to be cleaned.

Two saber-toothed blennies invade a cleaning station.

The fierce grouper
and other carnivorous
fishes spare the tiny
cleaner wrasses.

Dirty dancing

The saber-toothed blenny *(right, bottom)* imitates the wrasse *(right, top)* as part of its masquerade to grab a quick meal. To do so, it mimics the swimming style, or dance, of the wrasse. Then, as a deceived fish approaches to be cleaned, the blenny turns suddenly, bites a piece out of the victim's fin, and makes its escape. Other staples of the saber-toothed blenny's diet are bottom-dwelling worms and eggs spawned by other fishes.

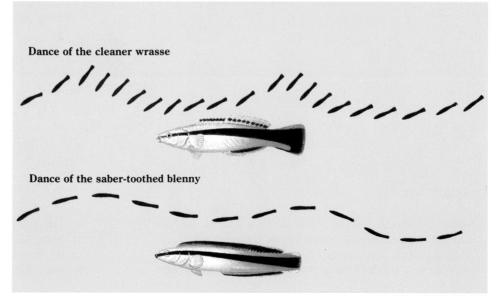

Dance of the cleaner wrasse

Dance of the saber-toothed blenny

How Do Crabs Use Camouflage?

Many crabs are fast enough to scuttle to safety when attacked. Others, lacking this protective speed, hide in crevices or under stones their entire lives.

Members of the decorator-crab and spider-crab family conceal themselves with camouflage. Their legs are covered with tough, curly bristles, which collect debris and small marine creatures until they are nearly covered. Some spider crabs, like the one shown here, complete the disguise by attaching bits of seaweed to their backs.

Hiding beneath a tarp of seaweed it stuck to its own back, an inch-wide spider crab *(left)* escapes the notice of a marauding grouper *(above)*.

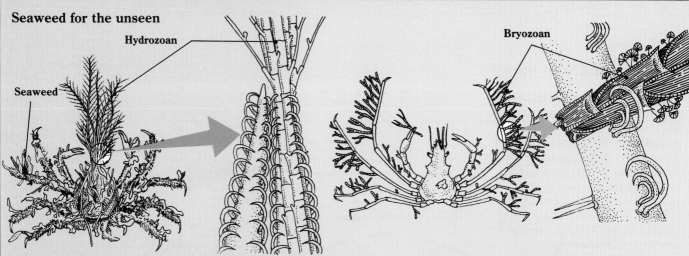

Seaweed for the unseen

Hydrozoan

Seaweed

Bryozoan

Some decorator crabs are densely covered with fine bristles, enabling them to encase themselves in seaweed and debris. Using its claws, a decorator crab may also attach sponges, anemones, or other tiny sea creatures to its shell and legs.

Other types of decorator or spider crabs have only a few hooklike bristles on their shells, so they cannot camouflage themselves as thoroughly.

Appropriate technology

In an aquarium experiment, the decorator crab at left was stripped of its natural seaweed camouflage and placed in a tank near a tangle of bright yarn. The crab promptly snipped off several strands of the yarn—the closest object—and stuck them to its shell. Because the yarn made the crab conspicuous, this action suggests that decorator crabs will use any available material to conceal themselves.

To disguise itself from predators, a decorator crab snips pieces of algae from the ocean floor, then fastens them to the top of its shell.

The evolution of camouflage

The decorator crab's ingenious camouflage technique may have evolved from its feeding habits. Eons ago, the crab was probably a simple benthic feeder—that is, it picked food off the seabed. Over time, the crab also began to eat seaweed and other objects that stuck to its shell and legs as it moved over the ocean bottom. Because these bits of seaweed made the crab less visible to predators, it may have next developed the practice of purposely sticking leftover food to its shell. Further evolution may have produced crabs that deliberately snipped seaweed and stuck it to their shells.

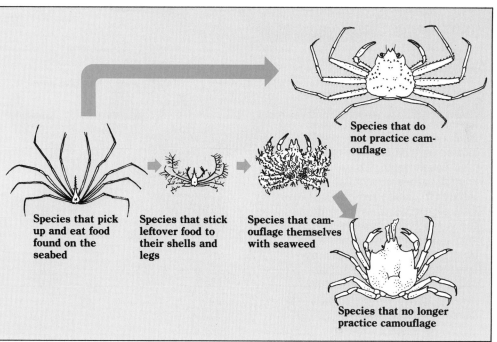

Species that do not practice camouflage

Species that pick up and eat food found on the seabed

Species that stick leftover food to their shells and legs

Species that camouflage themselves with seaweed

Species that no longer practice camouflage

How Does a Starfish Eat Its Prey?

Despite their graceful looks, starfish are aggressive, meat-eating predators. From the beach to the ocean depths, they kill and devour a remarkable number of their aquatic neighbors. Starfish are especially fond of bivalves, but they will also eat sea urchins, coral, some of the weaker crabs, and even dead fish.

To consume its prey, a typical starfish pushes its stomach out through its mouth and engulfs the food. The stomach may then be retracted, in which case the prey is swallowed whole and di-gested inside the starfish's body. In many species, however, digestion begins outside the body. Digestive enzymes reduce the meal to a thick paste, which is absorbed into the stomach along tracts lined with short, waving hairs called cilia. The stomach is then retracted.

Many marine creatures have developed special ways of fighting off or escaping from a starfish. Certain species of shrimps and crabs, for example, can snip off or wound the tube feet of an attacking starfish.

A scallop *(right)* escapes an attacking starfish *(below)* by blasting off from the seabed on a jet of water. The scallop creates the water jet by rapidly opening and closing its shell.

Under attack by a starfish, a keyhole limpet opens its shell and spreads out its white mantle, which can cut off the starfish's arm.

Anatomy of a starfish

Although some species of starfish have as many as 40 arms, most have just five. The arms radiate from a central disk. The mouth, located on the underside of this disk, has no teeth. To eat, the starfish pushes its stomach out through its mouth and envelops the food. Digestive juices break down the victim's nutrients, which are sent to the digestive glands to be absorbed. Some starfish can pry apart the two halves of a clam, insert their stomach through an opening as narrow as 1/200 inch, and digest the meat inside.

Anus — Vascular pore
Stomach — Skeletal plates
Gonad
Digestive gland — Tube feet

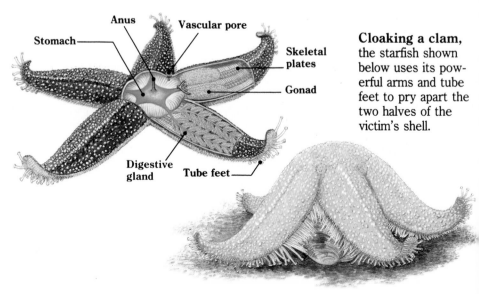

Cloaking a clam, the starfish shown below uses its powerful arms and tube feet to pry apart the two halves of the victim's shell.

One type of gastropod *(below)* is able to escape the clutches of a starfish by violently twisting its shell in one direction and then the other.

To repel an attacking starfish, a sea urchin flattens its spines, exposing barbed organs called pedicellariae *(right),* some of which contain a strong poison.

The surf clam *(right)* uses its powerful foot to spring to safety when it is threatened by a starfish.

Crabs and shrimps fight back

The crown of thorns, a starfish with up to 20 arms, lives among reefs, where it feeds on coral polyps—tiny, stationary, flowerlike animals that some early naturalists mistook for plants. When the starfish completes its grazing—it can consume the polyps in an area twice its size in one day—all that remains of the coral-building polyps is their white skeletons. Certain shrimps and crabs, however, protect the coral branches from this onslaught; as shown at right, they repulse the invading starfish by wounding its tube feet.

Why Does a Squid Squirt Ink?

To throw a predator off its trail, a squid often releases a dark, viscous liquid commonly referred to as ink. The ink coagulates in the water, forming a dense black cloud that decoys the pursuer long enough for the squid to change color and escape. Eventually, the cloud of ink disperses.

A squid's ink is produced in a special gland connected to its rectum. The ink consists primarily of melanin—the same substance that colors the hair and skin of humans and many other animals.

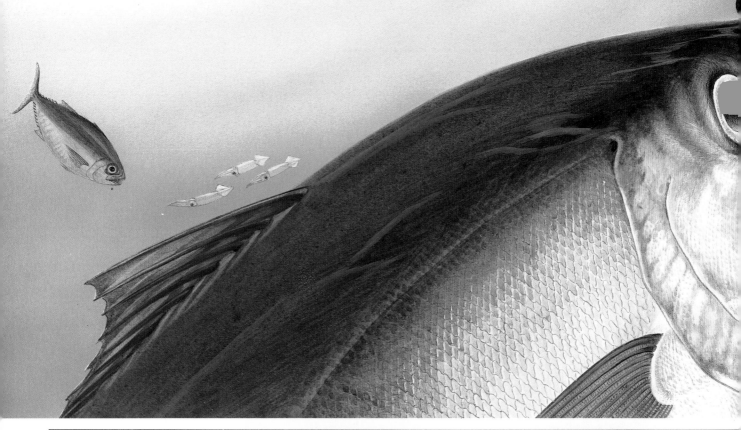

Veiled by visible ink

A cloud of black ink would be useless for protecting deepwater squid such as the one at right, which lives in the perpetual dark of the ocean depths. The deepwater squid therefore squirts luminous ink, whose eerie glow is generated by bacteria. The phosphorescent ink distracts potential predators long enough for the squid to escape.

Beset by a predatory tuna, a squid squirts out a pall of black ink. The cloud of ink—roughly the size and shape of the squid itself—acts as a decoy, diverting the attacker and giving the squid time to flee.

A different kind of ink

Like its 10-armed cousin the squid, the eight-armed octopus squirts ink when threatened. The octopus uses the liquid as a screen rather than a decoy; the ink is so thick it hides the octopus from view, allowing it to slink away unseen. The ink of the octopus may have other defensive uses as well: It is thought to contain an ingredient that numbs the senses of sight and smell in the moray eel and other fish that regularly prey on octopuses.

How Do Squid Change Their Markings?

Like a chameleon of the deep, the squid can change the color of its body markings in seconds. It does this not only for defense, but also as a prelude to communication and mating.

Tiny cells called chromatophores are the key to the squid's quick-change artistry. These red, black, yellow, and blue cells are arranged in thin layers of muscles under the skin. When the squid contracts the muscles, the chromatophores expand, making the squid darker or more colorful. When the squid relaxes the muscles, the chromatophores shrink, making the squid lighter or less colorful. By activating different combinations of these cells, the squid can produce a wide variety of body markings that help it survive.

By relaxing its muscles, the squid causes its color cells, or chromatophores, to condense. The squid's body becomes light or translucent.

A coat of many colors

The squid changes its body markings for many reasons. The first is to escape predators. But changes in color may also indicate excitement or a desire to communicate with other squid.

As shown at right, the color changes are not limited to a single shade that spreads over the entire body. The squid may suddenly display dark bands on both its tentacles and its mantle or on its mantle alone.

Some species alter their coloring to attract a mate. The male of some of these species turns his entire body black—except for his single testis, which remains a light color.

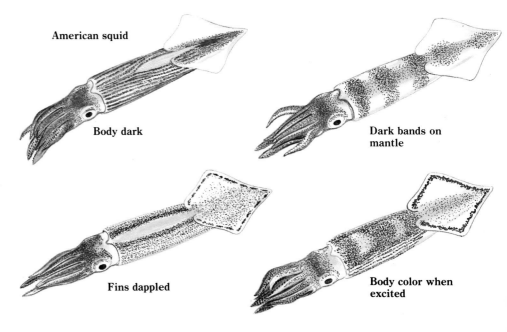

American squid

Body dark

Dark bands on mantle

Fins dappled

Body color when excited

With chromatophores of various colors arranged in layers under its skin, the squid can change its coloring simply by contracting certain muscles.

Dean of deep-sea disguise

Like the squid, the octopus has the power to vary its body coloring by relaxing or contracting its chromatophores. But its means of disguise do not stop there. A bottom-dwelling octopus can change the texture as well as the color of its skin; it can also alter the shape of its body. The bottom-dwelling octopus keeps a constant watch on its surroundings. When danger nears, it quickly makes itself look like the seafloor (*far right*). So expert is the octopus at using this camouflage that predators swimming nearby cannot distinguish the creature from the seabed on which it rests.

An octopus revealed

An octopus concealed

Do Fish Use Each Other for Protection?

Certain fish form a give-and-take partnership with another species. The clownfish, for example, lives peacefully among the poisonous tentacles of the sea anemone. The anemone protects the clownfish from possible attacks by predators; in return, the anemone benefits from morsels of food provided by the fish. This mutually beneficial living arrangement is known as symbiosis, or a symbiotic relationship.

Many fish, however, do not maintain such evenhanded relations with their fellow creatures. The fish shown here are examples of how some species exploit and even harm their hosts.

Young medusafish *(above)* live among the tentacles of a jellyfish, whose sting cannot hurt them. The medusafish may even nibble at the jellyfish's body.

These cardinalfish are both hosts and guests. Their bodies contain luminous bacteria, which make them glow. When threatened, they retreat to safety among the long, poisonous spines of the sea urchin.

The eel-like pearlfish hides during the day inside the body of a sea cucumber; it comes out to feed at night. An ungrateful guest, the young pearlfish frequently feeds on the sea cucumber's intestines.

Glossary

Aorta: The large artery that carries blood from the heart to smaller, branching arteries.

Artery: A thick-walled tube that carries oxygen-rich blood away from the heart and through the body.

Bioluminescence: The production and emission of light by a living organism.

Bivalve: A mollusk, such as a clam, whose shell has two halves, usually joined on one side by a hinge.

Cetacean: Belonging to the group of marine mammals that includes the whales and the porpoises.

Choroid: The membrane that lies between the retina and the sclera in the eyes of vertebrates.

Cold-blooded: Lacking the ability to regulate body temperature. The body temperature of a cold-blooded animal fluctuates with that of the surrounding air or water.

Cornea: The transparent outer layer of the eye, covering both the iris and the lens. The cornea protects the eye and helps focus light onto the retina.

Crustacean: Belonging to the group of mostly aquatic invertebrates that have segmented bodies, jointed limbs, and exterior skeletons. Shrimp, lobsters, barnacles, and crabs are all crustaceans.

Dermis: The inner layer of skin, beneath the epidermis.

Diadromous: Migrating between fresh water and salt water.

Echolocation: *See* Sonar.

Endolymphatic duct: A tube leading from the inner ear of a shark to a pore on the outside of its body.

Epidermis: The outermost layer of skin.

Estivate: To spend the summer in a dormant state.

Fascia: A sheet of connective tissue that binds together internal structures of the body.

Fin: An external membrane of an aquatic animal that is used to move or guide it. The **anal fin** is located behind the anus; the **caudal fin** is also called the tail fin; the **dorsal fin** is located on the back; **pectoral fins** are usually found on the sides, behind the gills; **pelvic fins** are located on the belly.

Fry: A recently hatched fish.

Gestation: The duration of a pregnancy.

Gills: Respiratory organs that act much like lungs. Gills are found in fish and in certain other aquatic animals, such as some crabs and amphibians.

Glycoprotein: An antifreezing protein found in the blood of fishes that inhabit polar seas.

Habitat: The local environment in which a plant or animal lives.

Hemoglobin: An iron-rich blood protein that carries oxygen.

Hermaphrodite: An animal with both male and female sexual organs.

Iris: The colored membrane, or diaphragm, that controls the amount of light entering the eye.

Lateral line: A shallow canal, located just beneath the skin on either side of a fish, that detects vibrations in the water.

Lens: The clear structure in the eye that focuses light.

Marine: A word used to describe organisms that live in the sea.

Metamorphosis: Any gradual but radical change that transforms an organism as it develops.

Mollusk: An invertebrate with a soft, unsegmented body, often enclosed in a hard shell; includes clams, snails, and octopuses.

Naris: The opening of the nose, or nasal passage; the plural is nares.

Nictitating membrane: A thin protective membrane that can be drawn across the eyeball of many animals.

Olfaction: The sense of smell.

Olfactory rosette: A circular structure in the nostril of a fish where sensory cells are located.

Operculum: The flap that covers the gills of a fish; also, the hard plate on the foot of a snail that is used to close the shell.

Osmosis: The movement of water or salt water through a porous membrane to equalize the concentration of salts on either side of the membrane.

Ossicle: A hard, bony structure found in the inner ear of vertebrates or in the soft tissues of jellyfish.

Oviparous: Producing eggs that hatch outside the mother's body.

Ovoviviparous: Producing young that hatch within the mother's body and are born live but receive no nourishment from the mother.

Pelagic: Living in the open ocean.

Photophore: A body organ that produces light.

Plankton: Tiny plants or animals that float and drift in water.

Planula: The juvenile, free-swimming stage of a jellyfish or related organism.

Polygamy: Having more than one mate.

Proboscis: An elongated tubular structure in the mouth region of an invertebrate.

Respiration: In animals, the process by which oxygen is inhaled and used by the body, and carbon dioxide is exhaled.

Rete mirabile: A network of blood vessels used to exchange gases or heat.

Retina: The rear interior surface of the eye, where light is focused.

Saccule: The smaller of two chambers in the inner ear.

Sclera: The outer lining of the eyeball.

Semicircular canal: One of three tubes in the inner ear that help a creature sense motion and keep its balance.

Septum: A membrane-like wall that separates tissues.

Siphon: A tubular organ, found in many mollusks and other invertebrates, used to draw fluids into the body or eject them.

Sonar: The detection of submerged objects from sound waves reflected back to the object emitting the sound; also known as echolocation. Some marine mammals use the technique to find prey and navigate.

Spawn: To produce or deposit fertilized eggs; also refers to the eggs themselves after they are spawned.

Species: A group of organisms that are similar to one another and can breed among themselves.

Spicule: A thin, hard, sharp-pointed structure that supports the body tissues of a sponge.

Subcutis: A deep layer of the dermis.

Swim bladder: The air- or oil-filled sac of a fish, used to help the fish rise, descend, or maintain neutral buoyancy.

Symbiosis: A close association between two organisms that often benefits each one.

Tapetum: A membrane that covers the choroid and retina of the eye.

Trochophore: The free-swimming larva of various marine invertebrates, including some mollusks and worms.

Univalve: A mollusk, such as a snail or a conch, with a one-piece shell.

Utricle: The larger of two chambers in the inner ear; the semicircular canals open into the utricle.

Vein: A tubular blood vessel that carries deoxygenated blood back to the heart or the gills.

Vitreous humor: A clear, jellylike filling that gives shape to the eyeball.

Viviparous: Producing young that develop and are nourished inside the mother's body.

Index

Staff for
UNDERSTANDING SCIENCE & NATURE

Assistant Managing Editor: Patricia Daniels
Editorial Directors: Allan Fallow, Karin Kinney
Writer: Mark Galan
Assistant Editor/Research: Elizabeth Thompson
Editorial Assistant: Louisa Potter
Production Manager: Prudence G. Harris
Senior Copy Coordinator: Juli Duncan
Production: Celia Beattie
Library: Louise D. Forstall
Computer Composition: Deborah G. Tait (Manager),
    Monika D. Thayer, Janet Barnes Syring, Lillian Daniels

Special Contributors, Text: John Clausen, Carol Dana,
    James Dawson, Marfé Ferguson-Delano, Barbara Mallen,
    Gina Maranto, Peter Pocock

Design/Illustration: Antonio Alcalá, Caroline Brock,
    Nicholas Fasciano, Stephen Wagner, David Neal Wiseman
Photography: Cover: David Doubilet. 1: Tom McHugh/Photo
    Researchers. 9: Gilbert van Rijckevorsel/Planet Earth
    Pictures, London
Index: Barbara L. Klein

Consultant:
    Stephen J. Walsh is a Research Fishery Biologist with the
    U.S. Fish & Wildlife Service in Gainesville, Florida. He also
    serves as a Courtesy Assistant Professor in the department
    of zoology at the University of Florida.

Library of Congress Cataloging-in-Publication Data
Underwater world.
    p.   cm. — (Understanding science & nature)
    Includes index.
    Summary: Questions and answers introduce the biology,
habitats, and behavior of aquatic animals, from luminous fish
to fur seals.
    ISBN 0-8094-9679-8 — ISBN 0-8094-9680-1 (lib. bdg.)
    1. Aquatic biology—Juvenile literature.
    [1. Aquatic biology—Miscellanea. 2. Marine animals—
Miscellanea. 3. Questions and answers.]
    I. Time-Life Books.   II. Series.
QH90.16.U53   1992
574—dc20   2507470
                                                 CIP
                                                 AC

TIME-LIFE for CHILDREN ®

Publisher: Robert H. Smith
Associate Publisher and Managing Editor: Neil Kagan
Assistant Managing Editor: Patricia Daniels
Editorial Directors: Jean Burke Crawford, Allan Fallow,
    Karin Kinney, Sara Mark, Elizabeth Ward
Director of Marketing: Margaret Mooney
Product Managers: Cassandra Ford, Shelley L. Schimkus
Director of Finance: Lisa Peterson
Financial Analyst: Patricia Vanderslice
Administrative Assistant: Barbara A. Jones
Special Contributor: Jacqueline A. Ball

Original English translation by International Editorial Services Inc./
C. E. Berry

Printed in Malaysia.
Published simultaneously in Canada.
Time Life Inc. is a wholly owned subsidiary of
THE TIME INC. BOOK COMPANY.
TIME LIFE is a trademark of Time Warner Inc. U.S.A.
For subscription information, call 1-800-621-7026.